Probabilistic Design for Optimization and Robustness for Engineers

Probabilistic Design for Optimization and Robustness for Engineers

Bryan Dodson

Executive Engineer, SKF, USA

Patrick C. Hammett

Lead Faculty Six Sigma Program, Integrative Systems & Design, College of Engineering, University of Michigan, Ann Arbor, USA

René Klerx

Principal Statistician, SKF, The Netherlands

WILEY

Library of Congress Cataloging-in-Publication Data

Dodson, Bryan, 1962–
  Probabilistic design for optimization and robustness for engineers / Bryan Dodson, Patrick C.
Hammett, René Klerx.
      pages cm
  Includes bibliographical references and index.
  ISBN 978-1-118-79619-1
  1. Industrial design–Statistical methods.   2. Reliability (Engineering)   3. Robust statistics.
  I. Hammett, Patrick C.   II. Klerx, René.   III. Title.
    TS171.9.D63 2014
    620′.00452–dc23

                                                                        2014013950

A catalogue record for this book is available from the British Library.

ISBN: 978-1-118-79619-1

Set in 10/12pt Times by Aptara Inc., New Delhi, India

Printed in the UK

Contents

Preface

Engineers spend years learning mathematical models to describe the behavior of systems. However, only a small portion of the engineering curriculum is dedicated to accounting for variation faced by product and process designers. Even here, the focus is usually limited to controlling manufacturing variation through tolerance analysis. Today, many engineering curricula offer elective courses in experimental design or robust design, but these courses focus more on system optimization and reducing variation in design through experimentation. This book presents the theory of modeling variation using physical models and presents methods for practical applications including making designs less sensitive to variation. This approach helps create designs that are easy to manufacture, with less design and manufacturing costs, and utilize more realistic tolerances. Methods are presented for determining nominal parameter settings that minimize output variation, determining the output variation caused by each input parameter, and minimizing total system costs, which includes the cost of non-conformance.

A challenge for this book is the lack of in-depth statistical training for many engineers. Many engineering curricula require a single course on probability or have no requirement at all. Stochastic modeling and optimization require some advanced statistical methods. Introductory chapters provide a logical roadmap to allow a complete understanding of the material without overwhelming the reader with excessive statistical rigor. Worked examples in the text are available on the Wiley website (www.wiley.com/go/robustness_for_engineers) along with animation software and computer-based exercises to aid understanding.

Acknowledgments

Paolo Re
Group Business Excellence
SKF

Rajeev Sundarraj
Graduate Student, Class of 2013
Industrial and Operations Engineering
University of Michigan, Ann Arbor

Silvio Vasconi
Regional Manager Engineering Consultancy Services
SKF

The authors wish to acknowledge the following individuals for their contribution in providing valuable input and industry examples.

Steven Geddes
Manufacturing Validation Solutions

Donald Lynch
SKF

Patrick Walsh
Manufacturing Validation Solutions

1

New product development process

1.1 Introduction

The development of new products is a major competitive issue as consumers continuously demand new and improved products. One outcome of this competitive landscape is the need for shorter product life cycles while still achieving ever increasing expectations for product quality and performance measures. This has required companies to significantly enhance their capabilities to better identify true customer wants, translate them into quantifiable product functional requirements, quickly develop, evaluate, and integrate new design concepts to meet them, and then effectively bring these concepts to market through new product offerings.

Several companies (e.g., Apple, General Electric (GE), Samsung, Toyota, General Motors (GM), Ford) have made great strides improving the effectiveness of new product development. For example, many companies have created processes to quickly gather voice of the customer information via surveys, customer clinics, or other sources. Samsung, for instance, has a well-designed system of scorecards and tool application checklists to manage risk and cycle time from the voice of the customer through the launch of products that meet customer and business process demands (Creveling et al., 2003). In addition, advances in computer simulation and modeling techniques permit manufacturers to evaluate many design concept alternatives, thereby resolving many potential problems at minimal costs. This also allows one to minimize assumptions and simplifications that reduce the accuracy of the answer (Tennant, 2002). Finally, even when there is a need to construct physical prototypes, the cost has been lowered through rapid prototyping processes.

Probabilistic Design for Optimization and Robustness for Engineers, First Edition.
Bryan Dodson, Patrick C. Hammett and René Klerx.
© 2014 John Wiley & Sons, Ltd. Published 2014 by John Wiley & Sons, Ltd.
Companion website: http://www.wiley.com/go/robustness_for_engineers

An interesting outcome of reducing the costs of data collection and analysis (for voice of the customer, simulation modeling, or physical testing) has been an increase in these activities. This has subsequently resulted in a deeper and broader understanding of customers and their interactions with products. This expanding knowledge base further allows a greater proliferation of product choices to satisfy increasingly diverse and sophisticated consumers.

Still, product development undoubtedly entails tremendous challenges. Many companies struggle with products that are slower to market than planned, fail to meet cost objectives, or are saddled with late design changes. Although no single recipe exists for product development success, one common thread is the ability to effectively integrate engineering resources within product and process design along with sales, marketing, manufacturing, and most importantly the end user.

Design for Six Sigma (DfSS) is a methodology that emphasizes the consideration of variability in the design process, resulting in products and processes that are insensitive to variation from manufacturing, the environment, and the consumer. The role of DfSS within new product development is to become an enabler of better integration of these resources to provide a deeper knowledge of product performance drivers and capabilities. An excellent example may be observed through GE, which has aligned the tools and best practices of DfSS within their product development process (Creveling et al., 2003). This chapter discusses the major phases of new product development with an emphasis on the roles engineers and DfSS resources play in effectively launching new products.

1.2   Phases of new product development

The time to develop a new product often depends on product complexity, which typically is a function of the technology readiness level (Assistant Secretary of Defense for Research and Engineering, 2011), the number of components, and the difficulties associated with manufacturing. In the case of an automobile, product development typically requires at least 2 years depending on the extent of the redesign. For example, if a manufacturer uses an existing powertrain and interior body frame, the development time may be reduced to less than 2 years. Product development times in aerospace industries typically range from 3 to 4 years, while the electronics industry is much faster with lead times of 6–12 months depending on the complexity of the product.

Although the total time for new product development will vary by design complexity and technological availability, the basic steps involved are common. Clark & Fujimoto (1991) and others (Tennant, 2002; Clausing, 1994) have provided basic descriptions of the product development process. The general phases (or steps) of new product development include concept development, product planning, product engineering design and verification, process engineering, and manufacturing validation as shown in Figure 1.1. The ideal situation for employing DfSS is to integrate it within these steps. To do so, one must acquire true customer needs and then apply the discipline of DfSS within the phases to efficiently transform customer needs into

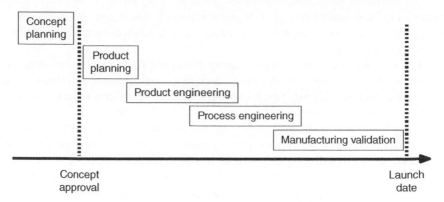

Figure 1.1   The phases of product development.

desirable products and services. DfSS and product development are complementary to each other and they can be implemented in parallel (Yang & El-Haik, 2003).

The following sections describe these phases in greater detail and discuss the roles of engineers and the integration of DfSS methods to improve their effectiveness.

1.2.1   Phase I—concept planning

During concept planning, manufacturers gather information on future market needs (voice of the customer), technological possibilities, and economic feasibility of various product designs. Many companies begin concept planning by expressing the character or image of their product in verbal, abstract terms using basic questions such as:

- Who shall use the product? (Target customers, cost of the product).

- What should the product do? (Performance and technical functions).

- What should the product have? (Appearance, packaging, key features, and options).

In defining a product concept, manufacturers often conduct three key assessments. These include assessing the voice of the customer, capabilities of the competition, and technological capabilities within the company.

The primary step in the development of a new product is the determination of the customer's wants and needs. Obtaining the voice of the customer traditionally has been the responsibility of Sales and Marketing who may conduct market studies, customer surveys, interviews, or use past sales data to identify market needs and trends. Although marketing is primarily responsible for customer research, under a DfSS framework, companies include more technical specialists such as product engineers in voice of the customer studies. The inclusion of technical specialists often accomplishes two objectives. First, product designers gain a better perspective of customer desires by mitigating the marketing filter. Second, technology specialists often are better suited to interpret emerging desires because of their deeper understanding of new technologies in development or existing ones in other industries that could be applied to their products.

To gain insight into consumer purchasing influences, Kano's method of analysis is a useful tool (Berger et al., 1993). Successful applications of Kano's methods require skill and experience. Translating customer wants and needs into product decisions remains a mix of art, science, and sometimes just good fortune.

To further assess the market, many companies conduct benchmarking studies of their competitors. Benchmarking is the continuous process of comparing one's own products, services, and processes against those of leading competitors. Although manufacturers typically benchmark direct competitors, they occasionally examine leaders in other industries. For example, car and bicycle manufacturers may benchmark airplane designs for ideas on how to make their products more aerodynamic, or for methods to improve internal processes.

To analyze complex products, today's manufacturers may even purchase their competitors' products and disassemble them down to evaluate the design. Here, companies are concerned with the inner workings of a product and how it is manufactured rather than its external appearance. Many companies set up "war rooms" where they make displays of competitor product components allowing internal engineers to review other designs and activate the creative process. In many cases, these war rooms provide a tremendous catalyst for making improvements. While one has to be careful to prevent benchmarking from leading to "look-alike" products, it can be a valuable tool to generate new ideas, which undoubtedly is necessary for continuous improvement of a product design.

The culmination of the concept and initial planning phase is often referred to as concept approval. This is an important date, because it typically is when financial resources are committed to bringing the product to market. While a company may reject a new product later in development, concept approval is generally *"when the clock starts ticking."*

1.2.2   Phase II—product planning

Once a concept is approved, a manufacturer must translate it into more concrete assumptions and detailed product specifications. In the language of DfSS, this involves the translation of customer requirements into product functional requirements, product attributes, and product features. This invariably consists of trade-offs between cost, functionality, and usability. Consider the design of an automotive body for a family sedan. Market studies may show that consumers want not only a strong rigid body for safety and handling, but also a vehicle with high gas mileage at a competitive cost. These few reasonable requests quickly create numerous design possibilities with each solution having its own set of advantages and disadvantages. For example, a manufacturer might choose to replace steel body panels with aluminum alloy panels because aluminum has a better strength to weight ratio. However, aluminum is generally more expensive than steel. It also can be more difficult to manufacture into certain shapes creating styling challenges. Under the DfSS framework, a manufacturer must establish a set of performance targets for the functional require-ments and then select a design which best meets them using a balanced scorecard approach (Yang & El-Haik, 2003).

Among the key activities that occur during product planning are styling, product architecture, and material and component selection. These activities are discussed in the following sections.

Styling and system architecture

Styling and system architecture are analogous to skin and bones. Styling represents the exterior appearance or exposed view of the product. Product architecture represents the structure and organization of internal components within a design system. In the design of a computer, stylists are concerned with the size, shape, and color of the monitor and computer box. Product architecture would be concerned with the positioning of the hard drive and external devices inside the computer box to improve functionality and lower manufacturing costs. Even in this simple example, the importance of integrating styling and architecture into a final design package becomes apparent. For example, in designing a tower computer box, the stylist might dictate the location and order of the external connections based on expected customer use, assuming the tower will be placed on the floor. Since USB connections are used more often than other devices, they may be placed closest to the top. In this example, stylist dictates the architecture.

Typically, companies do not use engineers to lead styling. For example, auto-motive manufacturers often utilize art and graphics specialists. These specialists are better trained at designing more appealing products. Still, while these non-engineers may drive styling, product design engineers remain essential to ensure product func-tionality and identify various manufacturing and cost limitations.

The authority of stylists or designers on the final product varies by company. Some companies rely heavily on designers and then expect engineers to determine how to make the design work. For others, product engineering may place a greater emphasis on how the product will function prior to determining how it looks ("form follows function"). Successful product developers clearly recognize that both styling and architecture must have similar levels of authority to effectively work together.

Material and component selection

Another critical role of engineers during new product development is material and component selection. New product development involves numerous choices between different types of material, new versus existing technology, in-house versus supplier parts, and various levels of sophistication for a particular technology. In all cases, engineers must consider the cost implications, effects on other components, and product concept objectives. Ultimately, companies must try to maximize value, where value represents the relationship between price and functionality (or quality); in other words, the amount a customer is willing to pay for a feature or function of a product.

During component selection, organizations identify advantages and limitations. For example, in the design of a mountain bike tire, engineers must decide how wide to make the tire while achieving weight targets and absorbing a specified level of road stress. One critical step in conducting such engineering is to understand the stresses that might incur under riding conditions. For example, a typical rider may only need to handle stresses incurred on gravel roads and jumps of less than one foot. If a manufacturer overdesigns their bicycle with excessively durable tires relative to the expectations of their target customers, they will produce an unnecessarily expensive product. While some customers may consistently ask for greater functionality, purchasing behaviors routinely suggest acceptance limits, often related to product prices.

1.2.3   Phase III—product engineering design and verification

Product engineering involves the execution of the product concept and planning phase. Product engineers construct detailed designs of the end product and its various components, including design verification. Here, many of the early engineering activities such as product architecture and component selection are reassessed during this phase as engineers add detail to the loose objectives identified in prior phases. Functional requirements are cascaded down from the system level to subsystems and eventually components. For example, the functional requirements of an automobile include safety and acceleration. Acceleration cascades down to the engine in terms of horsepower. Engine horsepower continues to cascade down to the piston and other components.

During process planning, a vehicle manufacturer may only decide between aluminum and steel for their doors. During product engineering, more detailed questions are addressed such as whether the door window should go directly into the roof panel or whether it goes into a header attached to the door itself. Furthermore, if an organization decides to use a door header, they then would need to determine whether the header should be a separate assembly attached to the lower door or integrated into the lower stamped door. Figure 1.2 illustrates three basic door design differences.

Once determining the basic system architecture, product engineering designs components and evaluates them against design criteria or functional requirements.

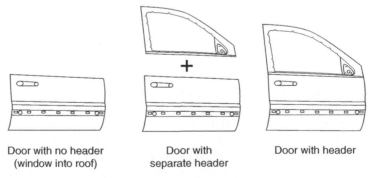

Door with no header   Door with   Door with header
(window into roof)   separate header

Figure 1.2   Door design alternatives.

Ideally this is done through engineering knowledge, including computer simulations. In cases where there are no engineering models, prototypes or replica are required for testing against design criteria or functional requirements. These criteria include both internal objectives and government standards such as safety and environmental regulations.

One way to consider multiple alternatives is through set-based concurrent engineering (Morgan & Liker, 2006). This approach involves considering a broad range of alternatives and systematically narrowing the sets to a final, often superior, choice. After finalizing the design plan, computerized drawings are created to convey the exact dimensions and requirements for each component. One important issue is to design interfaces that allow manufacturing to effectively assemble individual components. In developing drawings, product design engineers usually specify allowable variations (known as tolerances) for these interface dimensions in which the product design may vary and still be able to meet final product quality objectives. Considering more than one alternative also reduces risk when the technological readiness level is a concern.

To design a complex product, companies must develop various levels of specialization or rely on other organizations. In vehicle manufacturing, most companies divide their engineering groups by major subsystems such as body, chassis, electrical systems, and engine. Even within a major subsystem like body engineering, additional layers of specialists exist for internal and exterior body structures. Further specialization occurs at the working level where one engineer may focus on designing doors and another may specialize in hoods.

While this narrow specialization enhances engineering expertise, it also makes resource coordination and component design integration more difficult. Ultimately, organizations must constantly strive to balance the development of engineering specialists with cross-trained engineers to effectively integrate related subsystems. Toyota combines a strong functional organization (headed by general managers) with the deep specialization of a chief engineer (Morgan & Liker, 2006). This structure allows the chief engineer to focus on the customer and the integration of the overall product, whereas the general managers concentrate on their specialized systems and developing expertise among their engineers.

To enable coordination and integration, downstream resources such as process engineers and manufacturing personnel must have a channel of communication to provide insight into potential design problems. Poor integration often leads to late changes in designs. These engineering changes may result from lack of understanding of customer requirements, insufficient product knowledge, insufficient process knowledge, or errors of omission.

DfSS aims to mitigate the lack of understanding of customer requirements by more systematically gathering the voice of the customer and then translating this information into a set of comprehensive product design requirements with appropriate target and acceptance limits for functional performance measures. Ford (FMEA Handbook, 2004) and SKF (Re et al., 2014) cascade the requirements between system levels with the use of boundary diagrams. Boundary diagrams clearly define inputs, outputs, and responsibility for each level of a design.

Other late engineering changes may be related to insufficient product or process knowledge. These changes often result from skipping or compressing evaluation cycles due to pressures to reduce product development timing and costs. Organizations cannot test every possible occurrence that could lead to a product failure. Advancements in computer simulation and modeling are helping to mitigate this issue. Still, the creation of effective physical testing and experiments in the field of use along with the usage of methods like experimental design will continue to play a critical role in cases where engineering knowledge is lacking.

Another type of design error ("errors of omission") occurs when a product engineer misses a requirement or fails to resolve a historical problem. Repeating historical problems often is related to companies not effectively maintaining component design histories that categorize problems from prior models. As a result, design problems are repeated, especially if experienced engineers retire or change positions.

To reduce the errors of omission, design engineers must effectively communicate with both upstream functions (marketing and planning) and downstream functions such as process engineering (design of processes to build components) and manufacturing (physically making or assembly of components). Communicating with downstream development processes is particularly important because engineering changes usually increase in cost as the start of regular production approaches. Although all companies experience some engineering changes, the number and severity of these changes relative to product launch dates often separate the leading product developers from others. Developers that are not World Class have an increasing number of engineering changes culminating during validation and then spiking again after launch. This is illustrated in Figure 1.3. In contrast, world class developers typically identify problems earlier and resolve issues by the start of product introduction.

Ford (FMEA Handbook, 2004) and SKF (Re et al., 2014) use a chain of documents to store and reuse knowledge. The house of quality translates customer desires into engineering functions. These functions become the outputs of the boundary diagram. The boundary diagram is augmented with a parameter diagram, which lists the sources of variation. These sources of variation become failure causes in the design Failure

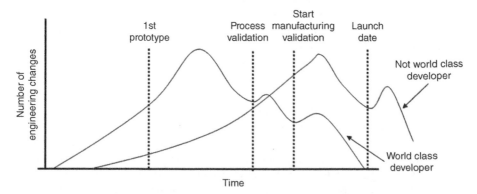

Figure 1.3   Engineering changes relative to start of production.

Mode and Effects Analysis (FMEA). The design controls in the design FMEA become the verification plan. The design FMEA also identifies potential characteristics critical to the design function and to safety. The critical characteristics are either confirmed or removed in the process FMEA based on the manufacturing capability. The critical characteristics information form summarizes the confirmed critical characteristics from the process FMEA, and becomes the foundation for the manufacturing control plan. This entire chain of documents exists at each level of the design, and is updated on a continuous basis. This ensures all design and manufacturing knowledge is retained in an easy-to-use format that is easily reused with future designs.

While requirements are cascaded from the system level to the component level, the opposite is true for verification. Component designs are verified first, then the subsystems that include the components are verified, and finally, the system design is verified. Computer models and engineering knowledge play an important role in verification. Verification is only required where there is a lack of engineering understanding, which includes any assumptions that may have been made when using engineering models or equations. Historically, a design, build, test cycle has been repeated until an acceptable performance level is achieved. The methods described in this text combined with computer engineering tools can be used to break this cycle. Ideally, an optimal design is obtained with a single iteration.

1.2.4   Phase IV—process engineering

During the process engineering phase, organizations translate component design information into manufacturing processes. Process engineering consists of numerous specialists in a variety of manufacturing fields such as casting, stamping, machining, injection molding, bonding, and welding. These activities might include designing cutting tools, new fixtures, and process control software, in addition to training workers and developing standard operating procedures. For example, if a vehicle manufacturer wants to produce a hood, they would need to construct new tooling. Tooling generally refers to the equipment that interfaces directly with the product. In the hood example, process engineers would take design drawings of the hood inner and outer components and develop stamping dies (tools) that produce these components using stamping presses. Process engineers might also design new measurement fixtures to check the quality of the stamped hood panels. Assembly process engineers then would be responsible for designing and developing hemming and any subsequent welding operations, which are used to join the hood inner panel to the hood outer panel and attach any additional components such as latches.

One difference between product and process engineering is organizations typically develop manufacturing processes for a longer life cycle. With the exception of certain tools like dies and molds, which are often designed specifically for a particular component, manufacturing processes are usually capable of producing a variety of products. For example, welding robots may simply require reprogramming if a manufacturer changes component designs. In fact, organizations purposely design flexibility into manufacturing processes so changes can be made to product designs without purchasing new tools or machines. In redesigning processes, organizations prefer to only change the exterior tooling rather than purchase new equipment. As a result, many organizations contract independent firms to design and build their manufacturing processes. The effect of this approach is that internal process engineers may serve as liaisons rather than process design specialists. Still, process engineers provide a critical link between the production factory, product designers, and external engineering resources.

The criticality of the process engineering function within the product development process often depends on the experience using a particular technology. For example, vehicle manufacturers have used resistance spot welding for years and are able to design welders to assemble components relatively quickly. In contrast, a manufacturer may decide to switch from resistance welding to laser welding to improve the quality of the weld. This switch likely will create unknown challenges requiring more process development time for testing, debugging, and validation.

In developing a process, manufacturers must assess the effects of various process input variables on product outputs. Product output characteristics, such as the length or diameter, typically are controlled by a number of input parameters specific to a particular manufacturing process. These parameters may be relatively simple to control like adjusting the machine cutting speed or more difficult such as controlling material flow during a metal forming operation. For more complex processes, establishing a relationship between inputs and output variables is substantially more difficult. Here, more sophistical analysis methods are needed. In addition, process engineers must also consider the robustness of the relationships and design processes accordingly.

Robustness of the manufacturing process makes the production more uniform despite variability (Clausing, 1994). This ultimately leads to both improved quality and lower manufacturing cost. A robust process is where an output variable is insensitive to the variation of an input variable over its operating range. The wider the robust range for an input variable, the easier it is to control during normal production. For every process, it is important to clearly document what will happen, how it will behave, how long will it take, and how much will it cost when various input adjustments are made to continuously improve robustness of the process (Nevins et al., 1989).

Product design engineers should consider the capability of the manufacturing process when creating designs. The following chapters describe how to predict output variation from knowledge of input variation. This knowledge can be used to ensure the planned manufacturing operations have adequate capability, and to trade tolerances between parameters to minimize product cost.

1.2.5   Phase V—manufacturing validation and ramp-up

After designing components and developing processes, organizations begin preparing for full-scale commercial production. One might think that at this point engineering is complete and meeting production launch date is the responsibility of the production department. This, however, is not the case. Product and process designs often are not completely finalized until after manufacturing validation, where problems may arise when assembling components at regular production line rates. These problems result from an inability to meet and verify original design requirements by the start of manufacturing validation, or the failure to build products under regular production conditions.

In the first case, some product launches must deal with components that have yet to meet all design specifications by the start of validation. This, in turn, hinders the ability for downstream functions to properly evaluate the assembly of a component to its mating parts. Of course, some components are late because of excessively tight specifications that are not necessary to meet final product customer and functional requirements. Here, the resolution requires better processes for establishing and linking requirements.

A second set of validation problems results from differences between making components in a controlled environment versus regular production conditions. During process engineering evaluations, manufacturers may hand load parts into machines. In regular production, however, these parts may need to be loaded automatically using conveyors and pick-and-place devices. In some cases, the effects of automation may result in unforeseen issues. In many cases, manufacturers will be unable to solve all of their problems by the product launch date, resulting in temporary extra inspection and repairs. As a result, the validation process may continue after the product launch date.

This period after the start of regular production is often known as ramp-up. Generally, this ramp-up period is considered as the time from production launch until a manufacturer is producing at full production line rate. Depending on the success of the development process, ramp-up may entail only a few days. In some cases, the multitude of problems results in ramp-up lasting for several months.

1.3   Patterns of new product development

It should be apparent from the prior discussion of the product development phases that the integration of resources and joint problem solving skills is critical to new product development. Product designers need to consider marketing concerns as well as the potential effect of designs on process engineering and manufacturing. Quality is a function of design.

The product development process historically has been done as a series of sequential activities by various product development functions. Figure 1.4 compares the level of activity of each function relative to launch dates for a sequential development approach. Here, process engineering does not start until after the design is nearly complete. This has led many to characterize this process as an "over-the-wall" process. In other words, rather than designers communicating closely with process engineers, designs are handed over, and designers begin work on the next project. The problem with this approach is that many design errors or problems are not uncovered until later in the process, where engineering change and rework costs are significantly higher.

In response to the limitations of sequential development, organizations have recognized the importance of overlapping development functions. Consider product and process engineering working in parallel. The parallel approach heightens the importance of coordination and communication. Product engineering must comprehend implications of their designs for manufacturability, and process engineering must

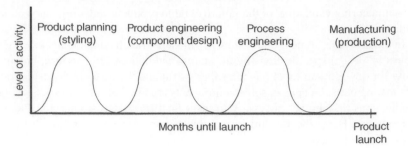

Figure 1.4   Sequential development (over-the-wall design).

clarify constraints and opportunities in the process and develop a good measure of flexibility to cope with the changes inherent in the product design process (Clark & Fujimoto, 1991). This greatly reduces lead time for development to better respond to changing market conditions. This approach is often known as concurrent or simultaneous development as illustrated in Figure 1.5. Under this approach, the development phases have significantly more overlap and thus require greater communication across resources.

While most organizations have some overlap in their development functions, the degree of overlap or concurrency may vary. For example, some organizations have their design engineers that not only solicit input from production departments, but also maintain authority and control over all design-related decisions. In contrast, some organizations give downstream functions such as process engineering and production, the authority to reject a particular design if they feel it cannot be manufactured effectively. Process designers at Toyota, for instance, use a "Lessons Learned Book." For instance, they have a Fender Die Design Book, which gives them a very detailed definition of what can be done (e.g., intervals of acceptable curvature radii for angles). Product design yields to these requirements. Of course, die design may develop a new technology or process to make the design feasible and revise the Lessons Learned Book (Ward et al., 1995). For another example, SKF encourages all engineers to understand the capability of manufacturing operations, and to release drawings with tolerances that can be achieved. SKF also models design outputs as a function of design inputs, and optimizes designs to be insensitive to variation in design inputs.

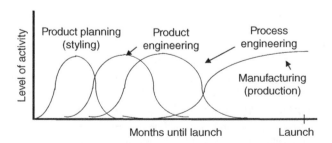

Figure 1.5   Concurrent product development.

One key to the integration of engineering resources is the organizational management structure. If new product managers have sufficient power and authority to make design changes and enhance communication channels, they can help insure better cooperation. However, if the new product manager is primarily a coordinator between product, process, and manufacturing departments, then new projects often have difficulty in making trade-offs and compromises between functions.

1.4   New product development and Design for Six Sigma

One way to enhance the capabilities of product development is to incorporate a DfSS approach or at least incorporate common DfSS tools into the product development process. DfSS involves a systematic approach to designing products to meet and exceed customer requirements while balancing internal business objectives for quality, cost, and timing. In short, it is a rigorous, systematic approach to develop higher value products in less time with less cost.

A key distinguishing features of DfSS are its focus on prevention of problems by designing optimal, robust processes that are less sensitive to typical operating conditions. Though often not obvious, an inherent lack of robustness in product design is a primary cause of manufacturing expenses. The "robustness" of products is more a function of good design, than manufacturing control, however, stringent the manufacturing process (Taguchi & Clausing, 1990). DfSS tools and methods are most effective during the phases from concept development to manufacturing validation. In contrast, once a product enters the later stages of manufacturing validation through regular production, problems become significantly more costly to correct, but of course they are much easier to identify. Here, the use of conventional quality problem solving and Six Sigma tools and methods such as the DMAIC process (Define, Measure, Analyze, Improve, Control) often is more effective for continuous improvement and achieving operational excellence. Figure 1.6 illustrates these differences.

As we discuss the link between DfSS and new product development, we may note that many of the individual tools and methods associated with it have been in existence long before DfSS became an established method embedded within new product development. In fact, the DfSS approach has a strong overlap with the push for "systems engineering." DfSS tools have been routinely employed, but not so stated (Vanek et al., 2008). Still, the adoption of DfSS methods provide a new opportunity to better link and apply many of these pre-existing tools to meet customer needs and reduce product costs.

1.4.1   DfSS core objectives

The core objective of DfSS is to create a more desirable design. This includes:

- Aligning products with customer requirements and desires.

- Reducing the product development cycle time with better knowledge reuse and by eliminating the build, test, fix cycle.

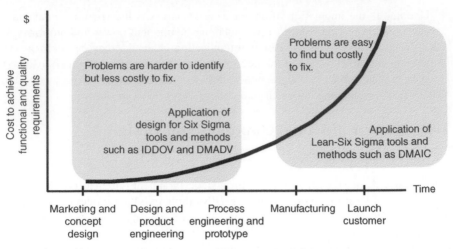

Figure 1.6   DMAIC versus DfSS.

- Designing products that are more robust to component, process, and user variation.

- Designing more reliable products.

- Designing products that are less costly and easier to produce.

- Enabling predictive versus reactive quality.

In terms of alignment, DfSS aims to give customers what they really want. This means not just what they say, but what they are willing to pay for. Alignment to customer requirements is also a never ending process as customers constantly raise their expectations or shift their preferences as new products become available.

Design for robustness involves delivering products and processes that are insensitive to the inevitable variability in manufacturing and use. While no product may be completely protected against all variation in processing or customer usage patterns, robustness may be quantified. Later in this book, we discuss several techniques to measure and quantify robustness.

While developing more robust and reliable products, which are essential, this must be done at a competitive cost. Critical to meeting cost targets is the ability to continuously develop products that are easier to manufacture than their predecessors. DfSS tools and methods may be used to evaluate and optimize design and process alternatives.

Finally, the adoption of DfSS stresses a shift from reactive to predictive quality. In a reactive world, requirements evolve often based on customer dissatisfaction or the inability of subsystems or components to meet cascaded specifications. Here, organizations employ build and test trials to determine a design solution, and then inspect in quality at the final product level. In contrast, a DfSS approach seeks to actively determine the voice of the customer and flow down requirements into a design solution. DfSS also stresses the use of simulation and modeling for initial evaluation and then the use of variation modeling for optimizing process parameters. In terms of quality, the objective is to design in quality and limit the amount of physical inspection for separating conforming from nonconforming product.

Given the broad objectives of DfSS, one must measure the overall quality of a design holistically. As such, we support the use of a total system desirability index that may include customer, functional, design, and processing requirements.

1.4.2   DfSS methodology

Several methods have been proposed to implement DfSS. For example, GM uses IDDOV (Identify-Define-Design-Optimize-Validate; Heincke, 2006); Ford prefers DCOV (Define-Characterize-Optimize-Validate; Stamatis 2004); while GE follows DMADV (Define-Measure-Analyze-Design-Verify; Snee & Hoerl, 2003). Other methods include DMEDI (Define-Measure-Explore-Develop-Implement; Costa, 2005), ICOV (Mader, 2003), and DCCDI (Define-Customer-Concept-Design-Implement; Tennant, 2002). Among these, we will illustrate DfSS using IDDOV.

The IDDOV process begins with the Identify and Define phase in alignment with the initial phases of new product development. Here, one gathers and prioritizes information on voice of the customer and functional requirements (design neutral requirements which quantify design performance and allow for creative solutions). These requirements flow down into the design phase where concepts are generated and selected. Design concept generation and selection in the context of DfSS starts at the system level, and is repeated at each level used within the end product. DfSS is also a useful tool when designing manufacturing processes.

Concept generation involves identifying new concepts and design alternatives, often by activating creativity through systematic or open innovation methods. Concept selection is done by comparing various alternatives against pre-established requirements and selecting the best overall option. Once a design concept is selected, computer-aided design and various other tools are used to create an actual product (either virtually or physically). At this point, the Optimize phase of DfSS begins to identify best settings for inputs using the robust design methods described in later chapters. This includes establishing target values and robust ranges for components and process settings along with the required tolerances and specifications to meet customer and functional requirements. Finally, the Verification & Validation phase is conducted to verify design intent, confirm that the product meets its requirements, and validate manufacturability.

A useful visual to show the integration of the IDDOV process with some common DfSS tools is the Systems V-diagram shown in Figure 1.7. The first mention of the V-diagram was found in Rook (1986), where it was introduced as a software project management tool illustrating the concept of verification of the process and products at established milestones (Kasser, 2010). The V-diagram was later introduced to the systems engineering community (Forsberg & Mooz, 1991). A key take away from this visual is that requirements flow down into a design solution via the IDD phases and then performance and quality measures flow up through the OV phases. A design

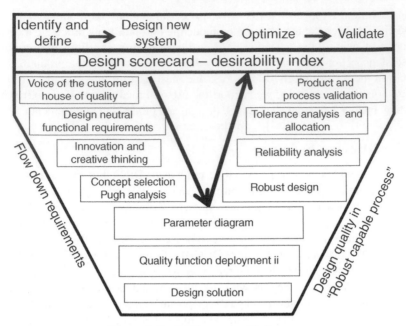

Figure 1.7   DfSS systems V-diagram.

scorecard with a desirability index may be used to measure the progress throughout (see Chapter 6 for additional details). It is important to note that the tools listed in the V-diagram are used for each level in the system.

1.4.3   Embedded DfSS

Similar to the conventional Six Sigma DMAIC approach, DfSS involves applying a structured, data driven methodology to solve high-impact problems and improve product performance. Among the quantifiable measures of DfSS successes may include:

- Shorter product development lead time.

- Lower total cost for product including engineering, materials, manufacturing, assembly, and shipping.

- Higher customer satisfaction ratings at launch and during the product life.

- Lower long-term production and operation costs.

- Higher reliability.

Unlike DMAIC, DfSS tools and methods often are embedded within an organization's existing new product development rather than applied as stand-alone improvement projects. Among "DfSS projects" led by Green or Black Belts, they often only contain either the IDD phases or IOV phases. As such, DfSS projects may be sub-classified as either IDD (Identify-Define-Design) or IOV (Identify Problem-Optimize-Validate) rather than full IDDOV. The power of DfSS is the organization of the tools into a coherent strategy that aligns with the product development process, not the individual tools themselves (Mader, 2003).

Several factors support the need to integrate DfSS within existing new product development. First, product development requires a substantial amount of company resources (both financial and resource time). Successful DfSS implementation should not be done in addition to normal activity but rather as an enabler to more effective product development. As the old adage says, "there's never enough time to do it right the first time, but there is always enough time to do it again." In fact, a popular view of DfSS is that it is not necessarily a new way of doing traditional development activities, but rather a more comprehensive, scientific, and data driven approach to make product decisions.

Since applying DfSS often occurs by incorporating the tools within an existing product development process, several challenges naturally arise. For instance, DfSS project applications typically span across multiple functions and organizations, thus the process owner may be unclear. Here, an organization must be careful to clearly identify roles and responsibilities related to a DfSS implementation. As noted by Soderborg (2004) of Ford Motor Company, DfSS implementation challenges often are more hindered by organizational and cultural change barriers than lack of technical skills to identify design improvements.

1.5   Summary

Developing an effective process to introduce new products clearly is a competitive weapon. A well-organized and efficient product development process is necessary to avoid cost overruns, late products, or product introductions with major quality problems. Although no single recipe exists for success, the ability of an organization to effectively integrate their engineering resources undoubtedly plays a central role. Engineers are trained to solve complex, practical problems. Rarely is this ability of greater value to a manufacturer than in the introduction of a new product. They simply need the tools and support to do so. The purpose of this book is to describe various methods and tools to more effectively develop new products.

Exercises

1   Choose a product or service recently launched in either your own organization or an organization closely related to your experience. Recall the process by which the new product or service was introduced. (For those with limited product development experience, look at media reviews for a new product.)

   (a) How well does this process align with the framework methodology discussed in this chapter? How does it differ?

(b) What is the level of overlap between the various stages of design (level of concurrent engineering)?

(c) Identify real or potential failures of the product/service and try to attribute the cause of this failure to a product development stage. Then, identify the extent to which a DfSS approach may have reduced or eliminated such failures.

2 Go to a nearby college or university and observe the various customers using "Backpacks." Note the different types of backpacks in use, and the classes or segments of customers using them, and the uses to which backpacks are put.

(a) Identify the product features involved in a typical backpack.

(b) Roughly segment the backpack types and also the customer base.

(c) What are the needs associated with each customer base in addition to the typical "basic" backpack? Which features delight and which annoy? Brainstorm and rank your own list. Identify new features that may overcome weaknesses for each customer type.

3 Consider the process of cooking a formal meal, such as for a party.

(a) How robust is this process? Recall experiences of unreliable delivery of a meal and determine the cause of failure.

(b) How much of the problem is due to human error and what fail-safe devices could be introduced to prevent this? What fail-safe devices are already in use?

(c) If you were completely redesigning the process, what changes would you implement to make the process more robust?

2

Statistical background for engineering design

This chapter presents a foundation of basic statistics required to understand the concepts covered in later chapters. Topics covered include statistical functions, expectation, statistical distributions, distribution parameters, and sampling.

2.1 Expectation

Probability density functions are grouped into two categories: discrete and continuous. The values a discrete density function can take on are limited to integers. For example, in five flips of a coin, the number of heads is limited to 0, 1, 2, 3, 4, or 5; the value 3.7 is nonsense. Continuous density functions can take on all values in a specific range. For example, the time to fail may be 1 hour, 1.01 hours, 1.001 hours, 1.0001 hours, etc.

To qualify as a density function, two criteria must be met:

(a) $f(x) \geq 0$

   for all values of x and

(b) $\int_{-\infty}^{\infty} f(x)\mathrm{d}x = 1$

   for continuous distributions, or for discrete distributions,

$$\sum_{n} f(x_n) = 1$$

   where the sum is taken over all possible values of n.

Probabilistic Design for Optimization and Robustness for Engineers, First Edition.
Bryan Dodson, Patrick C. Hammett and René Klerx.
© 2014 John Wiley & Sons, Ltd. Published 2014 by John Wiley & Sons, Ltd.
Companion website: http://www.wiley.com/go/robustness_for_engineers

The cumulative distribution function is the area under the probability density function to the left of the specified value, and represents the probability of x being less than a specific value, $P(X < x)$. For continuous distributions, the cumulative distribution function is defined as

$$F(x) = \int_{-\infty}^{x} f(\tau)d\tau \qquad (2.1)$$

The probability of x falling between two specified values, a and b, is the area under the probability density function from a to b, which is

$$P(a < X < b) = \int_{a}^{b} f(x)dx,$$

which is also equal to $F(b) - F(a)$. This is represented graphically in Figure 2.1.

Note that for continuous distributions, the probability of x being exactly equal to any specific value is zero. As seen in Figure 2.1, as b moves closer to a, the area (which is equal to the probability) decreases. When b is equal to a, the area is zero.

For discrete distributions, the probability of x being exactly equal to a specific value must be greater than zero for some values. For example, in four flips of a coin, where x represents the number of heads obtained, the probability of obtaining exactly zero heads is 0.0625, the probability of obtaining exactly one head is 0.25, etc. This is the probability density function, and is shown in Figure 2.2.

The cumulative distribution function for discrete distributions is

$$F(x) = \sum_{\tau \le x} f(\tau) \qquad (2.2)$$

The cumulative distribution function for the coin flipping example is given in Figure 2.3.

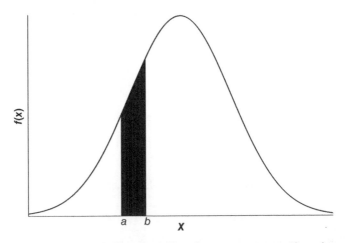

Figure 2.1   Probability of x falling between two specific values.

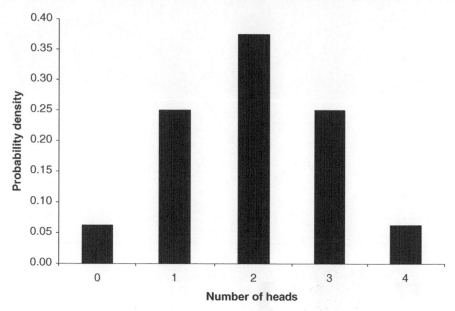

Figure 2.2   A discrete probability density function.

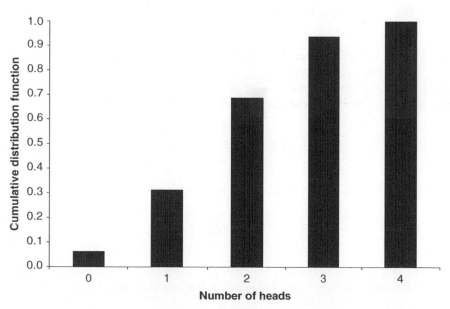

Figure 2.3   A discrete cumulative distribution function.

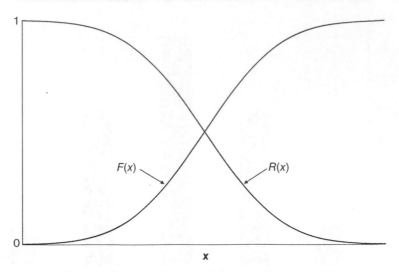

Figure 2.4   The relationship between R(x) and F(x).

The reliability function is the complement of the cumulative distribution function, $F(x)$, and is obtained from the probability density function using the relationship

$$R(x) = 1 - F(x) = 1 - \int_{-\infty}^{x} f(\tau)\mathrm{d}\tau = \int_{x}^{\infty} f(\tau)\mathrm{d}\tau \qquad (2.3)$$

The relationship between the reliability function and the cumulative distribution function is shown in Figure 2.4.

Distribution characteristics may be obtained from moment generating functions. If x is a continuous random variable, the nth moment about the origin is

$$E\left(x^{n}\right) = \int_{-\infty}^{\infty} x^{n}f(x)\mathrm{d}x \qquad (2.4)$$

If x is a discrete random variable, the nth moment about the origin is

$$E\left(x^{n}\right) = \sum_{x} x^{n}f(x) \qquad (2.5)$$

The first moment about the origin is the distribution mean or expected value of the distribution. For continuous distributions, the expected value is

$$E(x) = \int_{-\infty}^{\infty} xf(x)\mathrm{d}x = \mu \qquad (2.6)$$

The second moment about the mean is the distribution variance. For continuous distributions, the variance is

$$V(x) = E(x^2) - [E(x)]^2 = \int_{-\infty}^{\infty} x^2 f(x)dx - \mu^2 \tag{2.7}$$

The third moment about the mean is the skewness of the distribution. If a single peaked distribution has a long tail to the right, it is said to be skewed right. The fourth moment about the mean is kurtosis, which is a measure of the peakedness of a distribution.

Example 2.1

Given the probability density function

$$f(x) = ax, \ 0 \le x \le 10$$

(a) Determine the value of a that makes $f(x)$ a valid density function.

(b) Determine the mean and variance of the distribution.

(c) Derive an expression for the reliability function.

Solution

For the density function to be valid, the area under the curve must equal one. Thus,

$$\int_0^{10} (ax)dx = 1 \Rightarrow \frac{a}{2}\left(x^2\right)\Big|_0^{10} = 1 \Rightarrow \frac{a}{2}\left(10^2 - 0^2\right) = 1 \Rightarrow a = \frac{1}{50}$$

From Equation 2.6, the mean of this distribution is

$$E(x) = \int_0^{10} x\frac{x}{50}dx = \int_0^{10} \frac{x^2}{50}dx = \frac{1}{150}(10)^3 = 6.667$$

From Equation 2.7, the variance of this distribution is

$$V(x) = \int_0^{10} \frac{1}{50}x^3 dx - (6.667)^2 = \frac{1}{200}(10)^4 - 44.449 = 5.55$$

The reliability function is

$$R(x) = \int_x^{10} \frac{1}{50}\tau d\tau = 1 - \frac{x^2}{100}, \quad 0 \le x \le 10$$

2.2   Statistical distributions

Statistical distributions fall into two categories: modeling distributions and sampling distributions. Modeling distributions are used to describe physical phenomena such as resistance, power, viscosity, etc. Sampling distributions are used to construct confidence intervals and test hypotheses. Sampling distributions will not be covered in this chapter as they are not required later in the text.

2.2.1   Normal distribution

Whenever several random variables are added together, the resulting sum tends to normal regardless of the distribution of the variables being added. Mathematically, if

$$y = x_1 + x_2 + x_3 + \cdots + x_n \tag{2.8}$$

then the distribution of y becomes normal as n increases. If the random variables being summed are independent, the mean and variance of y are

$$\mu_y = \mu_{x_1} + \mu_{x_2} + \mu_{x_3} + \cdots + \mu_{x_n} \tag{2.9}$$

$$\sigma_y^2 = \sigma_{x_1}^2 + \sigma_{x_2}^2 + \sigma_{x_3}^2 + \cdots + \sigma_{x_n}^2 \tag{2.10}$$

When several random variables are averaged, the resulting average tends to normal regardless of the distribution of the variables being averaged. Mathematically, if

$$y = \frac{x_1 + x_2 + x_3 + \cdots + x_n}{n} \tag{2.11}$$

then the distribution of y becomes normal as n increases. If the random variables being averaged have the same mean and variance then the mean of y is equal to the mean of the individual variables being averaged, and the variance of y is

$$\sigma_y^2 = \frac{\sigma^2}{n} \tag{2.12}$$

where σ^2 is the variance of the individual variables being averaged.

The tendency of sums and averages to become normally distributed as the number of variables being summed or averaged becomes large is known as the *Central Limit Theorem* or the *Theory of Large Numbers*. For distributions with little skewness, summing or averaging as few as three or four variables will result in a normal distribution. For highly skewed distributions, more than 30 variables may have to be summed or averaged to obtain a normal distribution.

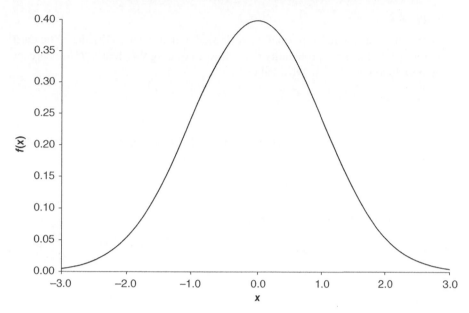

Figure 2.5   The standard normal probability density function.

The normal probability density function is

$$f(x) = \frac{1}{\sigma\sqrt{2\pi}} e^{-\frac{1}{2}\left(\frac{x-\mu}{\sigma}\right)^2}, \quad -\infty < x < \infty \qquad (2.13)$$

where μ is the mean and σ is the standard deviation.

The normal probability density function is symmetrical, as shown in Figure 2.5.

The density function shown in Figure 2.5 is the standard normal probability density function. The standard normal probability density function has a mean of zero and a standard deviation of one. The normal probability density function cannot be integrated implicitly. Because of this, historically, a transformation to the standard normal distribution is made, and the normal cumulative distribution function is read from a table. If x is a normal random variable, it can be transformed to standard normal using the expression

$$z = \frac{x - \mu}{\sigma} \qquad (2.14)$$

The transformation to standard normal is no longer required. Computers can easily compute the area beneath the normal probability density function.

Example 2.2[1]

The flow rate in a channel is normally distributed with a mean of 40 and a standard deviation of 6. What is the probability of the flow rate being less than 52? Five percent of the population is less than what value?

Solution

The probability of the flow rate being less than 52 can be found using Excel® with the function.[2]

```
=NORM.DIST(52,40,6,1) = 0.977
```

Five percent of the time the flow rate will be less than

```
=NORM.INV(0.05,40,6) = 30.13
```

Example 2.3[3]

A resistor is produced with an average of 20 ohms with a standard deviation of 0.5 ohms. If nine resistors are selected at random, what is the probability that the total resistance of the nine resistors is greater than 182? What is the probability that the average resistance of the nine resistors is less than 19.8?

Solution

The mean of the total resistance is $(9)(20) = 180$. The standard deviation of the total resistance is

$$\sigma = \sqrt{9\left(0.5^2\right)} = 1.5$$

Given a normal distribution with a mean of 180 and a standard deviation of 1.5, the probability of being greater than 182 is 0.091. This can be found using Excel® with the function.

```
=1-NORM.DIST(182,180,1.5,1) = 0.091
```

---

[1] This example is contained on the accompanying website. The file name is Normal_CDF.xls.

[2] The functions NORM.DIST and NORM.INV only work in Excel® version 2010 and later. Earlier versions use the functions NORMDIST and NORMINV.

[3] This example is contained on the accompanying website. The file name is Central_Limit_Theorem.xls.

The distribution of the average of nine resistors has a mean of 20 (the same as the average of an individual resistor), and a standard deviation of

$$\sigma = \frac{0.5}{\sqrt{9}} = 0.1667$$

Given a normal distribution with a mean of 20 and a standard deviation of 0.1667, the probability of being less than 19.8 is 0.115. This can be found using Excel® with the function.

```
=NORM.DIST(19.8,20,0.1667,1) = 0.115
```

2.2.2 Lognormal distribution

If a data set is known to follow a lognormal distribution, transforming the data by taking a logarithm yields a data set that is normally distributed. This is shown in Table 2.1.

The most common transformation is made by taking the natural logarithm, but any logarithm, such as base 10 or base 2, also yields a normal distribution. The remaining discussion will use the natural logarithm denoted as "ln."

When random variables are summed, as the sample size increases, the distribution of the sum becomes a normal distribution regardless of the distribution of the individuals. Since lognormal random variables are transformed to normal random variables by taking the logarithm, when random variables are multiplied (or divided), as the sample size increases, the distribution of the product becomes a lognormal distribution regardless of the distribution of the individuals. This is because the logarithm of the product of several variables is equal to the sum of the logarithms of the individuals. This is shown below.

$$y = x_1 x_2 x_3$$
$$\ln y = \ln x_1 + \ln x_2 + \ln x_3$$

Table 2.1   Transformation of lognormal data.

Lognormal Data	Normal Data
12	ln(12)
16	ln(16)
28	ln(28)
48	ln(48)
87	ln(87)
143	ln(143)

The lognormal probability density function is

$$f(x) = \frac{1}{x\sigma\sqrt{2\pi}} e^{-\frac{1}{2}\left(\frac{\ln x - \mu}{\sigma}\right)^2}, \quad x > 0 \tag{2.15}$$

where μ is the location parameter or log mean and σ is the scale parameter or log standard deviation.

The location parameter is the mean of the data set after transformation by taking the logarithm, and the scale parameter is the standard deviation of the data set after transformation.

Given the population mean, $E(x)$, and population variance, $V(x)$, the parameters of the lognormal distribution can be found with the following expressions.

$$\sigma = \sqrt{2 \ln [E(x)] - \ln \left[E^2(x) - V(x)\right]} \tag{2.16}$$

$$\mu = \ln [E(x)] - \frac{\sigma^2}{2} \tag{2.17}$$

where σ is the log standard deviation parameter and μ is the log mean parameter.

Example 2.4[4]

Given a lognormal distribution with a mean of 11.78 and a variance of 24.297, determine the parameters of the lognormal distribution.

Solution

The log standard deviation is

$$\sigma = \sqrt{2 \ln(11.78) - \ln \left[(11.78)^2 - 24.297\right]} = 0.439$$

The log mean is

$$\mu = \ln(11.78) - \frac{(0.439)^2}{2} = 2.37$$

The lognormal distribution takes on several shapes depending on the value of the log standard deviation. The lognormal distribution is skewed right, and the skewness increases as the value of σ increases. This is shown in Figure 2.6.

---

[4] This example is contained on the accompanying website. The file name is Lognormal_Parameters.xls.

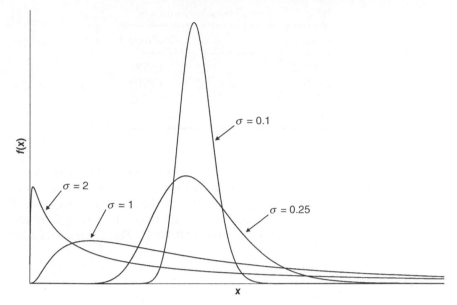

Figure 2.6   Lognormal probability density function.

The lognormal cumulative distribution function is

$$F(x) = \Phi\left(\frac{\ln x - \mu}{\sigma}\right) \tag{2.18}$$

where $\Phi(x)$ is the standard normal cumulative distribution function.

Example 2.5[5]

The data below are the power in watts for four circuits chosen at random. Since power is the result of voltage squared divided by resistance, it is expected to follow a lognormal distribution. Assuming a lognormal distribution, what percentage of the population of circuits has a power of less than 130 watts?

$$145.3 \quad 133.0 \quad 152.4 \quad 159.9$$

Solution

Table 2.2 shows this data and the transformation to normal.

---

[5] This example is contained on the accompanying website. The file name is Lognormal_Circuit_Prob ability.xls. Results shown in the text are found by carrying additional decimal points. Slight rounding errors will result by using the rounded ln(Power) values from Table 2.2.

Table 2.2   Transformation to normal.

Power	ln(Power)
145.3	4.9788
133.0	4.8903
152.4	5.0265
159.9	5.0745

The average of the power after transforming to normal by taking the natural logarithm is 4.9926, and the sample standard deviation of the transformed data is 0.07855. The percentage of circuits with power less than 130 is estimated as the probability of being less than the natural logarithm of 130 given a normal distribution with a mean of 4.993 and a standard deviation of 0.07855. This value can be found in Excel® with the function.[6]

```
=NORM.DIST(LN(130),4.993,0.07855,1) = 0.055
```

2.2.3   Weibull distribution

The Weibull distribution is one of the most flexible distributions available. It is capable of modeling data that are symmetrical, right skewed, and left skewed. It is commonly used to model systems with multiple components. The Weibull probability density function is

$$f(x) = \frac{\beta}{\theta} \left(\frac{x}{\theta} \right)^{(\beta-1)} e^{-\left[\left(\frac{t}{\theta}\right)\right]^{\beta}}, \quad x \geq 0 \tag{2.19}$$

where β is the shape parameter and θ is the scale parameter.

Given the population mean, $E(x)$, and the population variance, $V(x)$, the parameters of the Weibull distribution can be found with the following expressions.

$$E(x) = \theta \Gamma \left(1 + \frac{1}{\beta} \right) \tag{2.20}$$

$$V(x) = \theta^2 \left[\Gamma \left(1 + \frac{2}{\beta} \right) - \Gamma^2 \left(1 + \frac{1}{\beta} \right) \right] \tag{2.21}$$

---

[6] The function LOGNORM.DIST(130,4.993,0.07855,1) eliminates the need to transform the input to normal.

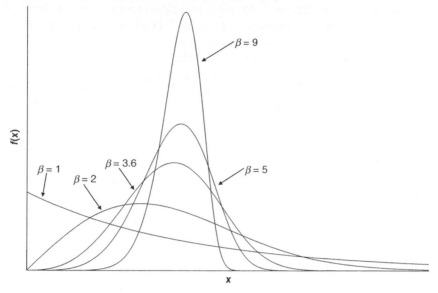

Figure 2.7   Weibull probability density functions with $\theta = 100$.

Example 2.6[7]

Determine the parameters of the Weibull distribution given a mean of 60.6 and a variance of 655.3.

Solution

The following two expressions can be solved simultaneously resulting in $\beta = 2.536$ and $\theta = 68.275$.

$$60.6 = \theta \, \Gamma \left(1 + \frac{1}{\beta} \right)$$

$$655.3 = \theta^2 \left[\Gamma \left(1 + \frac{2}{\beta} \right) - \Gamma^2 \left(1 + \frac{1}{\beta} \right) \right]$$

The shape parameter is what gives the Weibull distribution its flexibility. By changing the value of the shape parameter, the Weibull distribution can model a wide variety of data. If $\beta = 1$, the Weibull distribution is identical to the exponential distribution, if $\beta = 2$, the Weibull distribution is identical to the Rayleigh distribution; if β is between 3 and 4, the Weibull distribution approximates the normal distribution. The Weibull distribution also approximates the lognormal distribution for several values of β. A sample of the Weibull distribution's flexibility is shown in Figure 2.7.

---

[7] This example is contained on the accompanying website. The file name is Weibull_Parameters.xls.

The scale parameter determines the range of the distribution. The scale parameter is also known as the *characteristic life*. Regardless of the value of the shape parameter; 63.2% of all values fall below the characteristic life.

The Weibull cumulative distribution function is

$$F(x) = 1 - e^{-\left(\frac{x}{\theta}\right)\beta} \tag{2.22}$$

Example 2.7[8]

The surface finish of a bearing raceway follows the Weibull distribution with $\beta = 2$ and $\theta = 30$. If the upper specification for surface finish is 50, what percentage of the population will exceed this specification? At what upper specification level would 99% of the population meet specification?

Solution

The percentage of parts with a surface finish exceeding 50 is

$$R(50) = e^{\left[-\left(\frac{50}{30}\right)^2\right]} = 0.062$$

By manipulating the expression for reliability, 99% reliability is achieved with a surface finish of

$$x = \theta \left[- \ln R(x)\right]^{(1/\beta)}$$
$$x = (30) \left[- \ln(0.99)\right]^{1/2} = 3.01$$

2.2.4 Exponential distribution

The exponential distribution is used to model items with a constant occurrence rate. The exponential distribution is closely related to the Poisson distribution. If a random variable, x, is exponentially distributed, then the reciprocal of x, $y = 1/x$ follows a Poisson distribution. Likewise, if x is Poisson distributed, then $y = 1/x$ is exponentially distributed. Because of this behavior, the exponential distribution is sometimes used to model the mean time between occurrences, such as arrivals or failures, and the Poisson distribution is used to model occurrences per interval, such as arrivals, failures, or defects.

The exponential probability density function is

$$f(x) = \lambda e^{-\lambda x}, \quad x > 0 \tag{2.23}$$

where λ is the occurrence rate.

---

[8] This example is contained on the accompanying website. The file name is Weibull_Probability.xls.

The exponential probability density is also written as

$$f(x) = \frac{1}{\theta} e^{-\frac{x}{\theta}}, \quad x > 0 \tag{2.24}$$

where θ is the mean.

From the equations above, it can be seen that $\lambda = 1/\theta$. The variance of the exponential distribution is equal to the mean squared.

$$\sigma^2 = \theta^2 = \frac{1}{\lambda^2} \tag{2.25}$$

The exponential probability density function is shown in Figure 2.8.

The exponential cumulative distribution function is

$$F(x) = 1 - e^{-\frac{x}{\theta}} = 1 - e^{-\lambda x}, \quad x > 0 \tag{2.26}$$

The exponential distribution exhibits a lack of memory, that is, the probability of survival for a time interval, given survival to the beginning of the interval, is dependent only on the length of the interval, and not on the time of the start of the interval. For example, consider an item that has a mean time to fail of 150 hours that is exponentially distributed. The probability of surviving through the interval 0–20 hours is

$$R(20) = e^{-\frac{20}{150}} = 0.8751$$

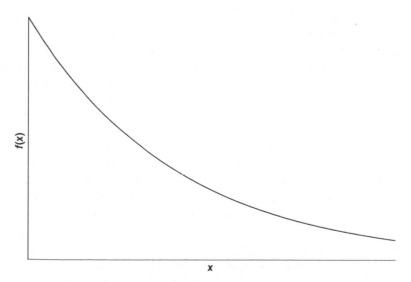

Figure 2.8   Exponential probability density function.

The probability of surviving the interval 100–120 is equal to

$$R(120, \text{ given survival to } t = 100) = \frac{R(120)}{R(100)} = \frac{e^{-\frac{120}{150}}}{e^{-\frac{100}{150}}} = \frac{0.4493}{0.5134} = 0.8751$$

Example 2.8[9]

If the length of university graduation speeches is exponentially distributed with a mean of 50 minutes, what is the probability that a randomly selected speech lasts less than 25 minutes? Fifty percent of all graduation speeches last less than what value?

Solution

The value of the cumulative distribution function at 25 is

$$F(25) = 1 - e^{-\left(\frac{25}{50}\right)} = 0.3935$$

Fifty percent of the time the length of a speech is less than $x = (50)$ $[-\ln(1 - 0.5)] = 34.7$ minutes.

2.3   Probability plotting

Probability plotting is a graphical method of parameter estimation. Probability plots are particularly useful for verifying the goodness-of-fit of the chosen distribution. A simple visual check of a probability plot provides a measure of the ability of a statistical distribution to model the data under consideration. In later chapters, statistical distributions will be chosen to compute probability based on knowledge of the physical behavior of the system being modeled. The validity of the chosen distribution can be assessed by constructing a probability plot using data generated from a Monte Carlo simulation[10] of the system.

To construct a probability plot, the cumulative distribution function for the assumed distribution is transformed to a linear expression and plotted. If the plotted points form a reasonably straight line, the assumed distribution is acceptable as a statistical model, and the slope and the intercept of the plot provide the information needed to estimate the parameters of the distribution of interest.

---

[9] This example is contained on the accompanying website. The file name is Exponential_ Probability.xls.

[10] Monte Carlo simulation is discussed in Chapter 4.

The cumulative distribution function, $F(x)$, is usually estimated from the median rank. The median rank[11] is found with the expression below by setting α to 0.5.

$$w_\alpha = \frac{\dfrac{j}{n-j+1}}{F_{1-\alpha,2(n-j+1),2j} + \dfrac{j}{n-j+1}} \tag{2.27}$$

where w_α is the $100(1 - \alpha)\%$ nonparametric confidence limit, j is the order of the data point, n is the total number of data points, and F_{α,v_1,v_2} is the percentile of the F distribution.

Benard's approximation for the median rank (below) is accurate to 1% with a sample of 5 and accurate to 0.1% with a sample of 50.

$$\hat{w}_{0.5} = \frac{j - 0.3}{n + 0.4} \tag{2.28}$$

By setting α to 5% and 95%, Equation 2.27 can be used to generate 5% and 95% ranks which are used for confidence limits.

2.3.1   Probability plotting—lognormal distribution

By rearranging the lognormal cumulative distribution function, a linear expression can be obtained.

$$\ln(x) = \mu + \sigma \, \Phi^{-1}\left[F(x)\right] \tag{2.29}$$

where $F(x)$ is the lognormal cumulative distribution function and $\Phi^{-1}(x)$ is the inverse of the standard normal cumulative distribution function.

By plotting $\ln(x)$ versus $\Phi^{-1}\left[F(x)\right]$, the resulting y-intercept provides an estimate for the log mean, μ, and the resulting slope provides an estimate for the log standard deviation, σ.

Example 2.9[12]

Given the data below, determine the parameters of the lognormal distribution using probability plotting.

14.3   8.1   9.8   19.2   7.5

---

[11] The accompanying website contains a spreadsheet named Ranks.xls that contains the rank function in the format "Ranks(alpha, total number of data points, order)."

[12] This example is contained on the accompanying website. The file name is Lognormal_Probability _Plot_Estimation.xls.

Table 2.3   Lognormal probability plotting data.

Order	x	$\ln(x)$	Median Rank $[F(x)]$	Standard Normal Inverse of $F(x)$
1	7.5	2.015	0.130	−1.128
2	8.1	2.092	0.315	−0.482
3	9.8	2.282	0.500	0.000
4	14.3	2.660	0.685	0.482
5	19.2	2.955	0.870	1.128

Solution

Table 2.3 is constructed to obtain the data for plotting. Note that the data must be ordered from smallest to largest. This data are plotted in Figure 2.9. The slope of this plot is 0.443, which is the estimated value of σ. The y-intercept is 2.401, which is the estimated value of μ. Since the plotted points form a reasonably straight line, there is no reason to disqualify the lognormal distribution as an adequate statistical model.

2.3.2   Probability plotting—normal distribution

By rearranging the normal cumulative distribution function, a linear expression can be obtained.

$$x = \mu + \sigma\,\Phi^{-1}\,[F(x)] \tag{2.30}$$

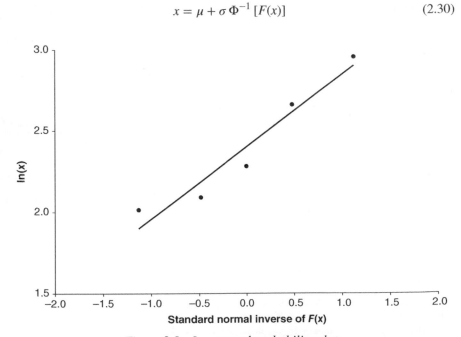

Figure 2.9   Lognormal probability plot.

Table 2.4   Normal probability plotting data.

Order	x	Median Rank $[F(x)]$	STD Normal Inverse of $F(x)$
1	25	0.130	−1.128
2	27	0.315	−0.482
3	29	0.500	0.000
4	31	0.685	0.482
5	32	0.870	1.128

where $F(x)$ is the normal cumulative distribution function and $\Phi^{-1}(x)$ is the inverse of the standard normal cumulative distribution function.

By plotting x versus $\Phi^{-1}[F(x)]$, the resulting y-intercept provides an estimate for the mean, μ, and the resulting slope provides an estimate for the population standard deviation, σ.

Example 2.10[13]

Given the data below, determine the parameters of the normal distribution using probability plotting.

$$31 \quad 27 \quad 29 \quad 32 \quad 25$$

Solution

Table 2.4 is constructed to obtain the data for plotting. Note that the data must be ordered from smallest to largest. These data are plotted in Figure 2.10. The slope of this plot is 3.264, which is the estimated value of σ. The y-intercept is 28.8, which is the estimated value of μ. Since the plotted points form a reasonably straight line, there is no reason to disqualify the normal distribution as an adequate statistical model.

2.3.3   Probability plotting—Weibull distribution

By taking the logarithm of the Weibull cumulative distribution function twice and rearranging,

$$\ln\left[\ln\left(\frac{1}{1-F(x)}\right)\right] = \beta \ln(x) - \beta \ln(\theta) \tag{2.31}$$

---

[13] This example is contained on the accompanying website. The file name is Normal_Probability_Plot _Estimation.xls.

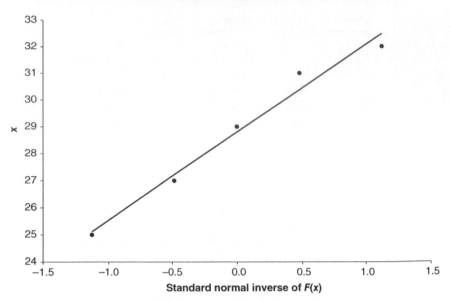

Figure 2.10   Normal probability plot.

By plotting $\ln[\ln(1/(1 - F(x)))]$ versus $\ln(x)$, and fitting a straight line to the points, the parameters of the Weibull distribution can be estimated. The slope of the plot provides an estimate of β, and the y-intercept can be used to estimate θ.

$$\theta = e^{-\left(\frac{y_0}{\beta}\right)} \tag{2.32}$$

Example 2.11[14]

Given the data below, determine the parameters of the Weibull distribution using probability plotting.

$$11 \quad 27 \quad 41 \quad 8 \quad 55 \quad 70$$

Solution

Table 2.5 is constructed to obtain the data for plotting. Note that the data must be ordered from smallest to largest. These data are plotted in Figure 2.11. The slope of this plot is 1.176, which is the estimated value of β. The y-intercept is -4.37105, which provides the estimated scale parameter for the Weibull distribution.

$$\theta = e^{-\left(\frac{-4.371}{1.176}\right)} = 41.1$$

---

[14] This example is contained on the accompanying website. The file name is Weibull_Probability_Plot _Estimation.xls.

Table 2.5   Weibull probability plotting data.

Order	x	$\ln(x)$	Median Rank $[F(x)]$	$\ln\left[\ln\left(\dfrac{1}{1-F(x)}\right)\right]$
1	8	2.079	0.109	−2.156
2	11	2.398	0.266	−1.175
3	27	3.296	0.422	−0.602
4	41	3.714	0.578	−0.147
5	55	4.007	0.734	0.282
6	70	4.248	0.891	0.794

Since the plotted points form a reasonably straight line, there is no reason to disqualify the Weibull distribution as an adequate statistical model.

2.3.4   Probability plotting—exponential distribution

By rearranging the exponential cumulative distribution function, a linear expression can be obtained.

$$\ln\left[\frac{1}{1-F(x)}\right] = \frac{x}{\theta} \tag{2.33}$$

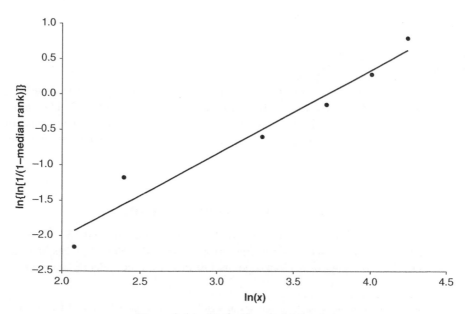

Figure 2.11   Weibull probability plot.

Table 2.6   Exponential probability plotting data.

Order	x	Median Rank $F(x)^a$	$\ln\left[\dfrac{1}{1-F(x)}\right]$
1	7.5	0.130	0.139
2	8.1	0.315	0.378
3	9.8	0.500	0.693
4	14.3	0.685	1.156
5	19.2	0.870	2.043

^{a}Benard's approximation was used to estimate the median rank in this example.

where $F(x)$ is the exponential cumulative distribution function and θ is the mean of the exponential distribution.

By plotting $\ln[1/(1-F(x))]$ versus x, the resulting slope from a regression line passing through the origin provides an estimate for the mean, θ. The regression line must pass through the origin because the exponential distribution is defined by a single parameter.

Example 2.12[15]

Given the data below, determine the parameters of the exponential distribution using probability plotting.

$$14.3 \quad 8.1 \quad 9.8 \quad 19.2 \quad 7.5$$

Solution

Table 2.6 is constructed to obtain the data for plotting. Note that the data must be ordered from smallest to largest. These data are plotted in Figure 2.12. The slope of a regression line through the origin for the data is 11.87, which is the estimated value of θ. Since the plotted points do not reasonably follow the regression line, the exponential distribution should not be used as a statistical model.

2.3.5   Probability plotting with confidence limits

Adding confidence limits to a probability plot defines the magnitude of the sampling error, and provides a visual guide for the error between the plotted points and the regression line. In general, if plotted points fall out of the confidence intervals or

---

[15] This example is contained on the accompanying website. The file name is Exponential_Probability _Plot_Estimation.xls.

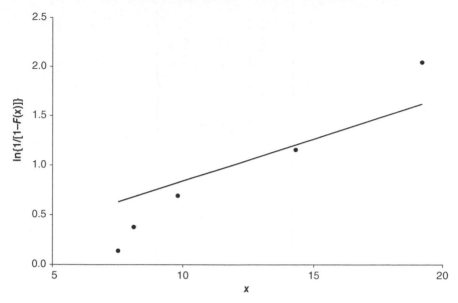

Figure 2.12   Exponential probability plot.

form a nonlinear pattern, then the distribution under investigation should be rejected as a statistical model.

Confidence limits are found using the rank distribution, and a 90% confidence interval is determined with 5% and 95% ranks—removing 5% of the distribution from each tail leaves 90% in the central part. The plotting points for the 5% and 95% ranks are not the same as for the median rank points. The plotting points are found at the intersection of the regression line, as any deviation of a point from the regression line is considered to be caused solely by sampling error if the distribution provides an adequate statistical model.

The plotting points for the confidence limits for the lognormal distribution are

$$\text{pp} = \mu + \sigma \, [F(x)]^* \tag{2.34}$$

(*Note that μ and σ are the log mean parameter and the log standard deviation parameter, respectively, not the population mean and standard deviation.)

The plotting points for the confidence limits for the normal distribution are

$$\text{pp} = \mu + \sigma \, [F(x)] \tag{2.35}$$

The plotting points for the confidence limits for the Weibull distribution are

$$\text{pp} = \theta \left[\ln \left(\frac{1}{1 - F(x)} \right) \right]^{1/\beta} \tag{2.36}$$

Table 2.7 Data for lognormal probability plot with confidence limits.

Order	x	ln(x)	Median Rank[a]	5% Rank	95% Rank	Standard Normal Inverse of Median Rank	5% Rank	95% Rank	Median Rank Plotting Position
1	7.5	2.015	0.129	0.010	0.451	−1.129	−2.319	−0.124	1.901
2	8.1	2.092	0.314	0.076	0.657	−0.485	−1.429	0.405	2.186
3	9.8	2.282	0.500	0.189	0.811	0.000	−0.881	0.881	2.401
4	14.3	2.660	0.686	0.343	0.924	0.485	−0.405	1.429	2.616
5	19.2	2.955	0.871	0.549	0.990	1.129	0.124	2.319	2.901

[a]Equation 2.27 was used to estimate the median rank for this example, not Benard's approximation.

Example 2.13[16]

Given the data below, create a lognormal probability plot with confidence limits.

$$14.3 \quad 8.1 \quad 9.8 \quad 19.2 \quad 7.5$$

Solution

Table 2.7 is constructed to obtain the data for plotting. Note that the data must be ordered from smallest to largest. The median rank is computed using the exact method as opposed to the approximation since the exact method is required for the 5% and 95% ranks. These data are plotted in Figure 2.13. The slope of this plot is 0.443, which is the estimated value of σ. The y-intercept is 2.401, which is the estimated value of μ. Since the plotted points form a reasonably straight line, there is no reason to disqualify the lognormal distribution as an adequate model for the data.

2.4   Summary

It is important to verify the correct distribution is being used. Improper distribution choice will result in errors. The errors may be small or large depending on the situation. Understand the statistical distributions, and verify the distribution fit and any assumptions before proceeding. In some cases, there may not be sufficient data to analytically determine the distribution fit. In these cases, the physics of the situation should be used to guide the distribution choice. If the data are symmetrical data, or the characteristic of interest results from a sum or an average, then a normal distribution is a good choice. The lognormal distribution is indicated when the characteristic of interest results from a combination of multiplication and division. The exponential distribution is used when there is a lack of memory, and the Weibull distribution is used for life estimates, and is often used as a catch all distribution because of its ability to model a wide range of data sets.

Probability plotting provides a visual check of fit, but a model is never proven to be valid, rather it is either rejected or there is not enough evidence to reject the model. With small sample sizes, the lack of enough data to reject a poor model often leads to several models that appear to be adequate. With very large data sets, it is common to reject all statistical models. In cases of small and large sample sizes, engineering judgment should be exercised. If a model is rejected due to a slight deviation on the right tail, it may still be valid if it is only used to compute probabilities on the left tail. In all cases, the physics of the situation should be considered. For example, the normal and lognormal distributions may both appear to represent a small data set equally well. If physics indicates right skewness, the normal distribution should be rejected even though the probability plot indicates a good fit.

---

[16] This example is contained on the accompanying website. The file name is Lognormal_Probability _Plot_Confidence_Limits.xls.

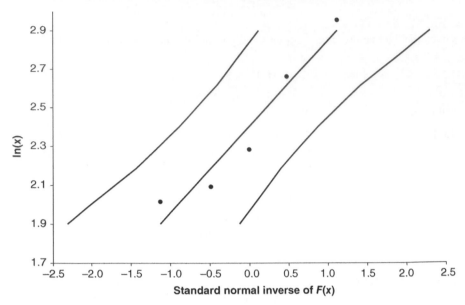

Figure 2.13   Lognormal probability plot with confidence limits.

Exercises

1 Given that a population of components has a normal time to fail distribution with a mean of 400 hours and a standard deviation of 60 hours, graph the probability density function, and the reliability function.

2 For the components in Exercise 2.1, if 200 components are tested for 430 hours, what is the expected number of survivors? What is the expected number of failures between time = 410 and time = 430?

3 For the components in Exercise 2.1, graphically show the probability of a failure before 425 hours of operation, 410 hours of operation, and from this graph, determine the probability of a failure in the interval from 410 hours of operation to 425 hours of operation.

4 Repeat Exercises 2.1 and 2.2 with the components following a lognormal time to fail with the same mean and standard deviation.

5 A population of components has a constant failure rate of 0.003 per hour. What is the probability of failure before 200 hours of operation?

6 Given the data below, determine the parameters of the normal distribution using probability plotting.

2   28   59   81

7 Given the data below, determine the parameters of the lognormal distribution using probability plotting. Include 5% and 95% confidence limits.

2   28   59   81

8 Given the data below, determine the parameters of the Weibull distribution using probability plotting.

2   28   59   81

9 Given the data below, determine the mean of the exponential distribution using probability plotting.

2   28   59   81

10 Given that the data below were randomly selected from a lognormal distribution, 68.75% is the probability of being less than what value?

5121.69   7.28   301.96   2.08   0.5   11.84

11 Given a population following a Weibull distribution with a shape parameter of 2.9 and a scale parameter of 82.55, 46.41% is the probability of being greater than what value?

12 Given a normally distributed population with a mean of 321.22 and a standard deviation of 33, what is the probability of a randomly selected item having a value between 347.8 and 430.9?

13 Given a normally distributed population with a mean of 344.96 and a standard deviation of 38.29, 21.91% is the probability of being less than what value?

14 Given a normally distributed population with a mean of 355.17 and a standard deviation of 35.96, what is the probability of the average of six randomly selected items being greater than 281.9?

15 Given that the data below were randomly selected from a lognormal distribution, what is the probability of a randomly selected item having a value less than 5.1?

14   18   29

3

Introduction to variation in engineering design

Engineers spend years studying physical models that are used to predict system behavior from input parameters. For example, an engineer may estimate the deflection in a beam based on the beam length, width, height, and modulus of elasticity. Typically, engineers are not taught to model the stochastic behavior of systems. Variability in design inputs is addressed through extreme techniques such as worst-case analysis and safety factors. This text presents the methods engineers may use to expand physical models to predict the variation in systems as a function of the variation of the input parameters. This prevents over designing and allows true system optimization including cost and performance of all functions.

3.1 Variation in engineering design

Variation is a part of the world we live in. The voltage measured at each of the outlets in your home will not be identical. The weight of each lug nut retaining the wheels on your car will vary. This variation makes life difficult for engineers. Engineers and process designers must understand and compensate for variation. The product engineers design may behave unexpectedly because the nominal values specified on drawings are difficult to achieve.

A parameter diagram (P-diagram), shown in Figure 3.1, is an ideal document for identifying sources of variation. A P-diagram shows the inputs, outputs, error states, control factors, and noise factors for a system. The design engineer has the ability to assign values and tolerances to control factors. Noise factors may add variation to system outputs, but the engineer cannot define targets or tolerances for these factors.

Probabilistic Design for Optimization and Robustness for Engineers, First Edition.
Bryan Dodson, Patrick C. Hammett and René Klerx.
© 2014 John Wiley & Sons, Ltd. Published 2014 by John Wiley & Sons, Ltd.
Companion website: http://www.wiley.com/go/robustness_for_engineers

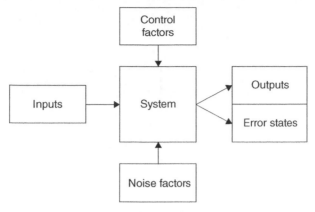

Figure 3.1   P-diagram.

Any noise factor that has a significant impact on the variation of any system output should be included in the model describing the system outputs. Noise factors are sub-divided into five categories:

1. Piece-to-piece variation—variation of parameters within engineering tolerances.

2. Degradation over time—changes in the nominal values of parameters over time. For example, the surface finish may become polished as an item wears in, or the resistance in a switch may increase as the switch ages.

3. Interactions with surrounding systems or components—often components work perfectly in a laboratory only to fail when assembled in a complete system. For example, an engine control module may fail because of electromagnetic compatibility issues when installed under the hood of an automobile.

4. Environment—variation in temperature, humidity, sand, salt, etc.

5. Customer use—variation in the number of duty cycles, loads, handling, etc.

3.2   Propagation of error

Variation in input variables propagates through the system resulting in variation of the system outputs. Consider a simple circuit consisting of a single power supply and a single resistor. The power in this circuit is

$$P = \frac{V^2}{R} \tag{3.1}$$

If the nominal voltage is 12 volts and the nominal resistance is 2 ohms, then the theoretical nominal power of the circuit is 72 watts. When this circuit is produced,

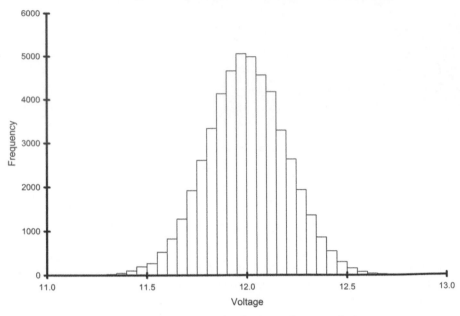

Figure 3.2   Histogram displaying voltage variation.

the voltage will not be exactly 12 volts, and the resistance will not be exactly 2 ohms, and thus the power will not be exactly 72 watts. If the voltage is normally distributed with a standard deviation of 0.2, and the resistance is normally distributed with a standard deviation of 0.1, then the standard deviation for the power would be 4.34[1]. This behavior is displayed in Figures 3.2, 3.3, and 3.4; which show histograms from a simulated production of 50 000 power supplies, resistors, and circuits.

Figure 3.4 shows the distribution of power is skewed right despite symmetrical distributions for voltage and resistance. Power variation is the result of variation in voltage and resistance propagating through the system. When a variable is a function of several variables, $y = f(x_1, x_2, \ldots, x_n)$, the procedure of determining the variability of y from knowledge of the variables, $x_1, x_2, \ldots, x_n$, is called propagation of error.

3.3   Protecting designs against variation

If the propagation of error is not modeled, engineers must find ways to protect designs against variation. Worst-case tolerancing and safety factors are common techniques used to protect against variation. An example of a safety factor is designing a balcony to support a load of 8000 kg when the balcony is required to support a load of 4000 kg.

---

[1] Details of how to estimate the output (power in this case) standard deviation are provided in the following chapters.

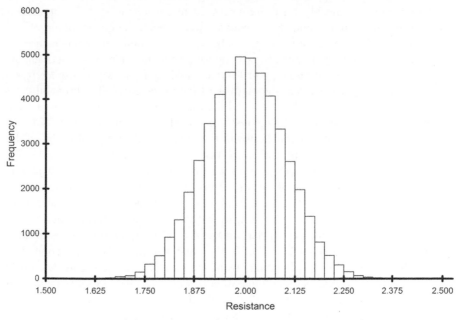

Figure 3.3   Histogram displaying resistance variation.

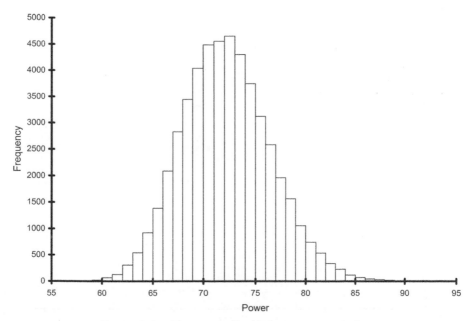

Figure 3.4   Histogram displaying power variation.

Worst-case tolerancing is a procedure used to ensure the system output will meet specifications if all inputs meet specifications. This technique is foolproof and efficient when the number of inputs is small, but as the number of inputs increases, the degree of over design required by worst-case tolerancing increases. This happens because the probability of all individual inputs aligning at the worst-case tolerance for a single system decreases as the number of inputs increases. The probability of two inputs both being on the lower side of the tolerance is 25% (50% squared), and the probability of five inputs all being on the lower side of the tolerance is 3.1% (50% raised to the fifth power). The probability of five inputs all aligning at the edge of the tolerance is essentially zero unless there is sorting in the production process.

Consider the simple case of building a wall with two bricks stacked vertically. If the wall height is required to be 20 cm tall ± 2 cm, then the worst-case tolerances of the individual bricks would be 10 ± 1 cm. If all bricks meet the tolerance requirement, then it is impossible for the wall height to fail to meet the tolerance requirement. Assume that the production process has marginal capability, is normally distributed, produces at an out of tolerance level of 0.27%, and that there is no inspection. The result is 0.27% of the bricks placed into walls are out of tolerance. Figure 3.5 shows a histogram from the results of building 50 000 walls with two bricks. Even though 270 (0.27% of 100 000 bricks) individual bricks did not meet tolerances, not a single wall consisting of two bricks failed to meet the tolerance requirements.

Now consider a wall with 10 bricks stacked vertically. If the wall height is required to be 100 cm tall ± 10 cm, then the worst-case tolerances for the individual bricks would again be 10 ± 1 cm.

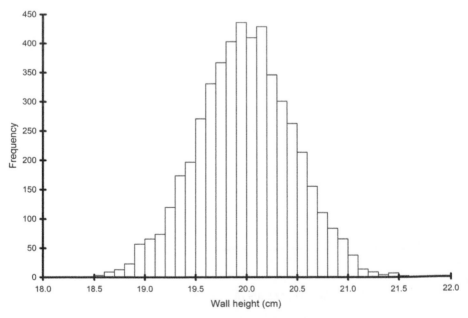

Figure 3.5   Histogram of wall height using two bricks.

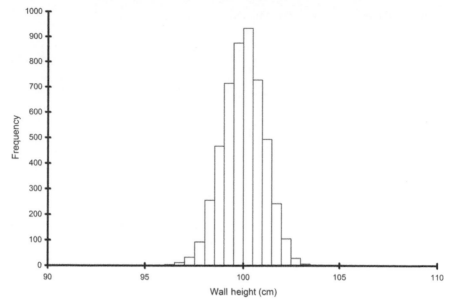

Figure 3.6   Histogram of wall height using 10 bricks.

Figure 3.6 shows a histogram from the results of building 50 000 walls with 10 bricks. The wall height consumes only 40% of the tolerance width (96–104). In this case, using the worst-case to determine the tolerance limits results in over design by more than a factor of two, and this would be exaggerated if the production process for the bricks were capable enough to produce all bricks within specifications. The individual brick tolerance width could be doubled without producing any walls that do not meet the required height tolerance. Engineers need a better method of predicting system variation to allow better input tolerancing and ultimately less costly and better performing designs.

The remainder of this chapter presents the methods for modeling propagation of error for simple systems.

3.4   Estimates of means and variances of functions of several variables

The sum or difference of normally distributed variables is also normally distributed. If x_1 and x_2 are normally distributed with means μ_{x1} and μ_{x2}, and variances $\sigma_{x_1}^2$ and $\sigma_{x_2}^2$, then $y = x_1 + x_2$ is normally distributed with a mean of

$$E(y) = \mu_{x_1} + \mu_{x_2} \qquad (3.2)$$

and a variance of

$$V(y) = \sigma_{x_1}^2 + \sigma_{x_2}^2 + 2\rho\sigma_{x_1}\sigma_{x_2} \qquad (3.3)$$

where ρ is the correlation coefficient, and is defined as

$$\rho = \frac{\text{cov}\left(x_1, x_2\right)}{\sigma_{x_1}, \sigma_{x_2}} \tag{3.4}$$

where $\text{cov}\left(x_1, x_2\right)$ is the covariance between variables x_1 and x_2, and is defined as

$$\text{cov}(x_1, x_2) = \int_{-\infty}^{\infty} \int_{-\infty}^{\infty} \left(x_1 - \mu_{x_1}\right) \left(x_2 - \mu_{x_2}\right) f\left(x_1, x_2\right) dx_1 dx_2 \tag{3.5}$$

The value, $y = x_1 - x_2$, is also normally distributed with a mean of

$$E(y) = \mu_{x_1} - \mu_{x_2} \tag{3.6}$$

and a variance of

$$V(y) = \sigma_{x_1}^2 + \sigma_{x_2}^2 - 2\rho\sigma_{x_1}\sigma_{x_2} \tag{3.7}$$

Note that Equations 3.3 and 3.7 are identical when the correlation coefficient is zero (when the variables are independent).

One of the most commonly used statistical principles is the theory of large numbers, also known as the central limit theorem. Simply stated, if

$$y = x_1 + x_2 + \cdots + x_n$$

where $x_1, x_2, \ldots, x_n$ are independent random variables, then, regardless of the distribution of $x_1, x_2, \ldots, x_n$, y will become a normally distributed as n becomes large. In this case, the mean of y is

$$E(y) = E(x_1) + E(x_2) + \cdots + E(x_n) \tag{3.8}$$

If $x_1, x_2, \ldots, x_n$ are independent, the variance of y is

$$V(y) = V(x_1) + V(x_2) + \cdots + V(x_n) \tag{3.9}$$

Figures 3.7 and 3.8 show the phenomenon of the sum of independent random variables tending to normal as the number of variables being summed increases. A distribution that has a shape much different from normal, the exponential distribution, becomes normal as n increases. Figure 3.7 shows the exponential distribution and the sum of two independent, exponentially distributed variables.

Figure 3.8 displays the sum of five, and the sum of ten, independent, exponentially distributed variables.

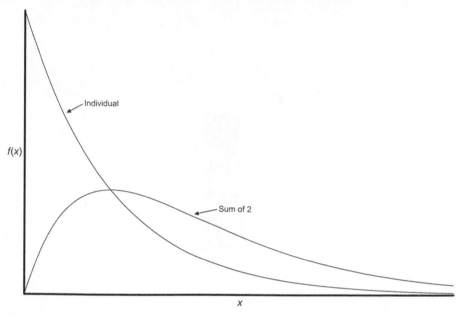

Figure 3.7   Exponential distribution and the sum of two exponentially distributed variables.

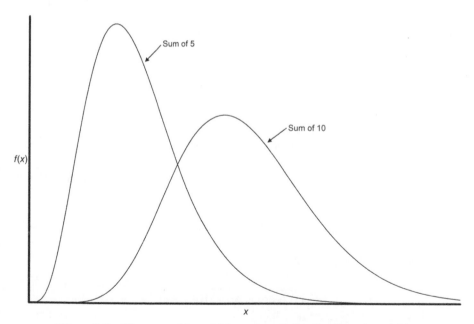

Figure 3.8   The sum of 5 and 10 exponentially distributed variables.

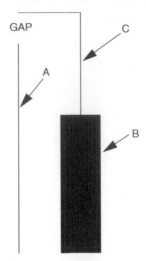

Figure 3.9   Sample fixture.

Example 3.1[2]

Referring to Figure 3.9, the sample fixture consists of three parts; the vertical bar to the left (part A), with a mean length of 10 cm and a standard deviation of 0.02 cm, the solid block (part B), with a mean of length of 7 cm and a standard deviation of 0.01 cm, and the vertical bar on top of the block (part C) with a mean length of 4 cm and a standard deviation of 0.03 cm. Determine the expected length and the standard deviation of the identified gap.

Solution

The mean of the gap is

$$E(\text{gap}) = 7 + 4 - 10 = 1$$

There is no reason to believe the length of these parts is not independent, thus, the variance of the gap is

$$V(\text{gap}) = (0.02)^2 + (0.01)^2 + (0.03)^2 = 0.0014$$

The standard deviation of the gap is

$$\sigma = \sqrt{V(\text{gap})} = \sqrt{0.0014} = 0.0374$$

---

[2] This example is contained on the accompanying website. The spreadsheet file is named "Chapter3.xls."

Example 3.2[3]

Consider the wall example discussed earlier. Ten bricks are stacked vertically, and the wall height is required to be 100 cm tall ± 10 cm. Determine the tolerances of the individual bricks.

Solution

If the tolerance limits are at least five standard deviations away from the average, a production process will generate less than one part per million out of tolerance. Thus, 10 cm is equal to five standard deviations of the wall height. The standard deviation of the wall is

$$\sigma_{\text{wall}} = \sqrt{10 \left(\sigma^2_{\text{brick}} \right)}$$

The required standard deviation for a brick is

$$5\sigma_{\text{brick}} = \sqrt{\frac{10^2}{10}} \Rightarrow \sigma_{\text{brick}} = 0.632$$

Setting the tolerances for the individual bricks at five standard deviations from the average, the brick tolerance is

$$10 \pm (5)(0.632) = 10 \pm 3.16$$

Note that this tolerance is more than three times as large as the worst-case tolerance, which should result in a substantial savings. Of course, this tolerance is only valid with a stable process that has no sorting.

Table 3.1 gives the mean and variance for simple functions of one or two variables.

Example 3.3[4]

Determine the expected area and the variance of the area of a rectangle with sides x and y. The variable x is normally distributed with a mean of 20 and a standard deviation of 0.4; the variable y is normally distributed with a mean of 60 and a standard deviation of 0.8.

Solution

From Table 3.1, the expected value of the area, $z = xy$ is

$$E(\text{area}) = (20)(60) = 1200$$

---

[3] This example is contained on the accompanying website. The spreadsheet file is named "Chapter3.xls."

[4] This example is contained on the accompanying website. The spreadsheet file is named "Chapter3.xls."

Table 3.1   Means and variances of functions of one or two variables.

Function	Mean	Variance
$y = ax$	$E(y) = aE(x)$	$V(y) = a^2 V(x)$
$y = a + x$	$E(y) = a + E(x)$	$V(y) = V(x)$
$y = x_1 + x_2$	$E(y) = E(x_1) + E(x_2)$	$V(y) = V(x_1) + V(x_2)^a$
$y = x_1 - x_2$	$E(y) = E(x_1) - E(x_2)$	$V(y) = V(x_1) + V(x_2)^a$
$y = x^2$	$E(y) = [E(x)]^2 + V(x)$	$V(y) = E(x^4) - [E(x^2) + V(x)]^2$
$y = x_1 x_2$	$E(y) = E(x_1)E(x_2)^a$	$V(y) = V(x_1)V(x_2) + V(x_2)[E(x_1)]^2$
		$\qquad + V(x_1)[E(x_2)]^{2a}$
$y = \dfrac{x_1}{x_2}$	$E(y) = E(x_1)E\left(\dfrac{1}{x_2}\right)$	$V(y) = \left[V(x_1) + E^2(x_1)\right] E\left(\dfrac{1}{x_2^2}\right)$
	$\qquad + \left(\dfrac{x_1}{x_2^3}\right) V(x_2)$	$\qquad - E^2(x_1)E^2\left(\dfrac{1}{x_2}\right)$

$^a x_1$ and x_2 are independent.

From Table 3.1, the variance of the area is

$$V(\text{area}) = (0.4^2)(0.8^2) + (0.8^2)(20^2) + (0.4^2)(60^2) = 832.1$$

While Table 3.1 gives distribution characteristics for transformed random variables, it is also useful to obtain the probability density function for the transformed variable. For the case of a single random variable, the probability density function for the transformed variable is

$$g(y) = f[w(y)]\left|\frac{dw}{dy}\right| \qquad (3.10)$$

where $f(x)$ is the probability density function for x, $y = u(x)$ is the transformation function, and $x = w(y)$ is the inverse of the transformation function, and has only one root.

Example 3.4

The random variable x is exponentially distributed with an occurrence rate of λ. Determine the probability density function for the variable $y = 4x + 12$.

Solution

For this example, the transformation function is $y = 4x + 12$. Thus, the inverse transformation function is

$$x = \frac{y - 12}{4}$$

The first derivative of the inverse transformation function is

$$\frac{dx}{dy} = \frac{1}{4}$$

The exponential probability density function is

$$f(x) = \lambda e^{-\lambda x}, \; x > 0$$

Substituting into Equation 3.10 gives

$$g(y) = \left(\lambda e^{-\lambda(y-12)/4} \right) \left(\frac{1}{4} \right), y > 12$$

The acceptable range for x is 0 to infinity, after transformation, when $x = 0$, $y = 12$, thus, the range of y is 12 to infinity.

If the inverse of the transformation function has more than one root, the probability density function for the transformed variable is

$$g(y) = \sum_{i=1}^{n} f\left[w_i(y) \right] \left| \frac{dw_i}{dy} \right| \tag{3.11}$$

where n is the number of roots of the inverse transformation function.

Table 3.2 contains some transformations of functions involving one variable.

Now consider the case where the random variables x_1 and x_2 have a joint probability distribution of $f(x_1, x_2)$ and are transformed to two new random variables, $y_1 = u_1(x_1, x_2)$ and $y_2 = u_2(x_1, x_2)$. If the inverse of the transformation functions, $x_1 = w_1(y_1, y_2)$ and $x_2 = w_2(y_1, y_2)$, have single roots, the joint probability density function for y_1 and y_2 is

$$g(y_1, y_2) = f\left[w_1(y_1, y_2), w_2(y_1, y_2) \right] |J| \tag{3.12}$$

where J is the Jacobian and is defined as the determinant of the partial derivatives

$$J = \begin{vmatrix} \partial x_1/\partial y_1 & \partial x_1/\partial y_2 \\ \partial x_2/\partial y_1 & \partial x_2/\partial y_2 \end{vmatrix}$$

In many cases, the transformation is to a single variable, y_1. In this case, simply choose a dummy variable for y_2. When choosing this dummy variable, be sure that y_1 and y_2 are independent; this will simplify the calculations because the joint probability density function for the two variables will be the product of the individual probability density functions.

Some transformations for functions of more than one random variable are given in Table 3.3.

Table 3.2   Transformations involving one random variable.

Transformation Function	Resulting Probability Density Function
$y = a + bx$	$g(y) = \left\| \dfrac{1}{b} \right\| f\left(\dfrac{y-a}{b} \right)$
$y = \dfrac{1}{x}$	$g(y) = \dfrac{1}{y^2} f\left(\dfrac{1}{y} \right)$
$y = e^x$	$g(y) = \left\| \dfrac{1}{y} \right\| f(\ln y)$
$y = \ln x$	$g(y) = e^y f(e^y)$
$y = x^2$	$g(y) = \dfrac{1}{2\sqrt{y}} \left[f\left(\sqrt{y} \right) + f\left(-\sqrt{y} \right) \right]$
$y = \sqrt{x}$	$g(y) = 2\,\|y\| f(y^2), \quad y > 0$
$y = x^{1/n}$	$g(y) = \left\| n y^{n-1} \right\| f(y^n), \quad y > 0$
$y = x^n$	$g(y) = \left\| \dfrac{1}{n} y^{(1/n)-1} \right\| \displaystyle\sum_{i=1}^{n} f(x_i)$
	$x_1, x_2, \ldots, x_n$ are the roots of $x = y^{1/n}$
$y = e^{ax}$	$g(y) = \left\| \dfrac{1}{ay} \right\| f\left[\ln\left(y^{1/a} \right) \right]$

$f(x)$ is the probability density function for x.

Table 3.3   Transformations for functions of more than one random variable.

Transformation Function	Resulting Probability Density Function
$y = x_1 + x_2$	$g(y) = \int f_{x_1}(y - z) f_{x_2}(z) \mathrm{d}z$
$y = x_1 x_2$	$g(y) = \int \left\| \dfrac{1}{z} f_{x_1}\left(\dfrac{y}{z} \right) \right\| f_{x_2}(z)\, \mathrm{d}z$
$y = \dfrac{x_1}{x_2}$	$g(y) = \int \left\| \dfrac{z}{y^2} f_{x_1}(z) f_{x_2}\left(\dfrac{z}{y} \right) \right\| \mathrm{d}z$
$y = x_1^2 + x_2^2$	$g(y) = \int \left\| \dfrac{1}{2\sqrt{y - z^2}} f_{x_1}(z) f_{x_2}\left(\sqrt{y - z^2} \right) \right\| \mathrm{d}z$
$y = \sqrt{x_1^2 + x_2^2}$	$g(y) = \int \left\| \dfrac{1}{2\sqrt{y^2 - z^2}} f_{x_1}(z) f_{x_2}\left(\sqrt{y^2 - z^2} \right) \right\| \mathrm{d}z$

f_{x_1} and f_{x_2} are the density functions for the random variables x_1 and x_2.

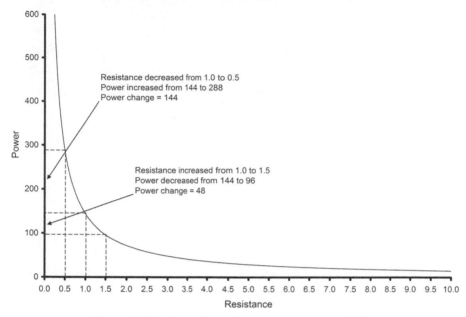

Figure 3.10   Power bias due to resistance variability.

3.5   Statistical bias

In nonlinear systems, the variability in input variables produces a statistical bias in the theoretical system outputs. Figure 3.10 shows the power in a circuit as a function of resistance with a 12-volt power supply. If the nominal resistance is 1 ohm, the power will be 144 watts. If the resistance varies by 0.5 ohms to the low side, the power increases to 288 watts. If the resistance varies by 0.5 ohms to the high side, the power reduces to 96 watts. Note that identical variation to the low and high side in resistance results in nonsymmetrical changes in power. The nonsymmetrical changes in power are the cause of the bias between the average of produced parts and the theoretical expected value.

Now consider the weight of a pipe as a function of length. This is a linear system, and the change in weight when length varies to the low side is exactly the same as the change in weight when length varies to the high side. This system will have no statistical bias. This is shown in Figure 3.11.

3.6   Robustness

In addition to causing a bias, the variance of a nonlinear system is dependent on the nominal value. It can be seen from Figure 3.10 that if the nominal resistance is increased, the variation of power will be reduced when the variation of the resistance is held constant. This is not true for linear systems. As seen in Figure 3.11, the

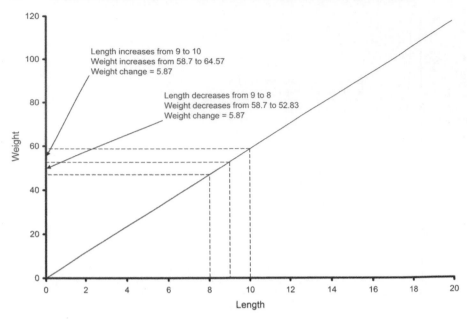

Figure 3.11   Weight bias due to length variability.

variability of weight is independent of the nominal length value. If the length varies by ±1, the weight will vary by ±5.87 regardless of the nominal value of length.

Consider the distance a projectile travels in a vacuum as a function of angle. With gravity and velocity held constant, the distance the projectile travels is shown in Figure 3.12.

Figure 3.12 shows that by choosing a nominal angle of 45°, the variability of the distance the projectile travels is considerably less than if a nominal angle of 15° were chosen. For this design problem, choosing a 45° nominal angle is the most robust solution. The variability of the output (distance) has been made insensitive to the variability of the input (angle).

3.7   Summary

Designing robust systems involves understanding how input variability propagates through the system resulting in system variability. Robustness is achieved by understanding the effect of input nominal values as well as input variability on system variability and how the system variability can be minimized. In addition, the impact of the variability of each input can be quantified, so important characteristics can be controlled and inputs with a small impact on the system can be targeted for cost savings. This chapter presented the methods for propagation of error and estimating statistical bias in simple systems, but methods are needed for the more complex systems faced by engineers. Methods for more complex systems are presented in the following chapters.

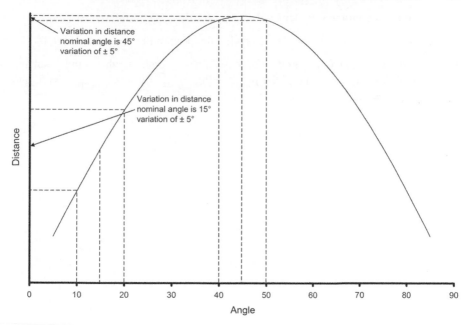

Figure 3.12   Variation of projectile distance traveled as a function of nominal angle.

Exercises

1 Give an example of a system that has a constant output variation over the entire range of input values.

2 Give an example of a system with an output variance that changes as the inputs change.

3 Determine the value of x that minimizes the output variance for the system below.

$$y = 12.4 \left[\cos \left(\frac{x + 340}{2.7\pi} \right) \right], 0 < x < 20$$

4 The mean of x_1 is 8.2 and the standard deviation of x_1 is 0.2. The mean of x_2 is 4.7 and the standard deviation of x_2 is 0.5. The correlation coefficient between x_1 and x_2 is 0.88. Determine the standard deviation of $y = x_1 + x_2$.

5 Determine the mean of the system given below $\mu_V = 12$, $\sigma_V = 0.1$, $\mu_R = 3$, and $\sigma_R = 0.2$.

$$I = \frac{V}{R}$$

6 Determine the standard deviation for the system in Exercise 3.5.

7 Determine the probability density function for the system in Exercise 3.5.

8 Given x is normally distributed, determine the probability distribution for $1/x$.

9 Given force is equal to the product of mass and acceleration, what is the mean and standard deviation of force given $\mu_{mass} = 20$, $\sigma_{mass} = 0.4$, $\mu_{acceleration} = 5$, and $\sigma_{acceleration} = 0.2$.

4

Monte Carlo simulation

As seen in Chapter 3, modeling the propagation of variability in a system can be mathematically complex. Monte Carlo simulation is a relatively easy method for predicting the variation and bias in a system. The procedure for performing a Monte Carlo simulation is:

1. define the system to be simulated,

2. determine the variation behavior of the inputs to the system,

3. for each input to the system, select a random number representative of the input's variation,

4. compute the system output(s), and

5. repeat the steps 3 and 4 until steady state is achieved.

4.1 Determining variation of the inputs

The variation of the inputs is determined by selecting a probability density function that adequately represents the behavior of the input and then determining the parameters of the distribution.

There are several options available for determining the variation of the inputs.

1. Use the tolerance window and assume a distribution and a process capability index (C_p) of 1.33.[1] Consider a 100-ohm 5% resistor. The tolerance window

---

[1] A C_{pk} of 1.33 is the generally accepted minimum capability index for a process. A C_{pk} of 1.67 (the average is 5 standard deviations from the nearest tolerance limit) may be used for parameters produced with superior capability.

Probabilistic Design for Optimization and Robustness for Engineers, First Edition.
Bryan Dodson, Patrick C. Hammett and René Klerx.
© 2014 John Wiley & Sons, Ltd. Published 2014 by John Wiley & Sons, Ltd.
Companion website: http://www.wiley.com/go/robustness_for_engineers

is 100 ohms $\pm$ 5 ohms. A C_{pk} of 1.33 is equivalent to $\pm$ 4 standard deviations resulting in a standard deviation of 1.25 (5/4) ohms. Since resistance varies in a symmetrical pattern, the normal distribution can be assumed. The result is a normal distribution with a mean of 100 and a standard deviation of 1.25. For characteristics that are not symmetrical, a lognormal or Weibull distribution may be used. For example, roundness is bounded at zero and is likely to be skewed to the right, so a lognormal distribution would be logical.

2. Request statistical data from the supplier or your manufacturing location. If your company is a large enough customer, a supplier will supply data from end-of-line tests. When requesting the data, ensure the samples are collected over a long enough time period to include all sources of variation including tooling changes, equipment maintenance, and multiple batches of raw material.

3. Collect your own statistical data. Consider a small company that will be using grease in a product being designed, and the viscosity of the grease will have a significant impact on friction. Enough data could be collected to fit a statistical distribution by periodically purchasing grease and measuring the viscosity. The purchases would need to be made over a long enough time period to ensure all sources of variation are included. Purchasing 30-50 packets of grease in a short period of time from the same location would underestimate the viscosity variability.

When determining input variation, a parameter diagram (P-diagram) may be useful. A P-diagram attempts to list all sources of variation. The variation may come from any of the following sources:

- piece-to-piece variation,
- degradation over time,
- environmental differences,
- customer usage, and
- system interactions.

For example, suppose the variance of the diameter of a pipe is needed, and a sample of 100 pipes is taken. The variance is estimated from sample measurements taken in a laboratory at room temperature. When the pipe is used, if the pipe is exposed to the environment, the variance of the diameter will have been underestimated because the variation from temperature fluctuations will have been neglected.

4.2   Random number generators

The heart of any simulation is the generation of random numbers. Random numbers from specific distributions are generated by transforming random numbers from the unit uniform distribution. The unit uniform distribution is uniformly distributed from zero to one as shown in Figure 4.1.

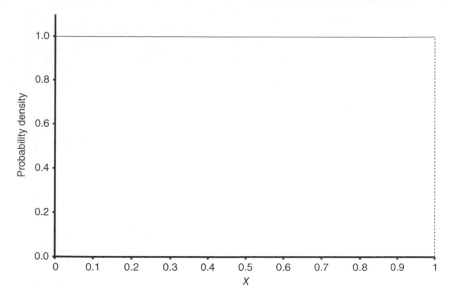

Figure 4.1   Unit uniform probability density function.

Virtually, all programming languages, as well as electronic spreadsheets, include a unit uniform random number generator. Technically, these unit uniform random number generators are pseudo-random number generators, as the algorithms used to generate them take away a small portion of the randomness. Nevertheless, these algorithms are extremely efficient and for all practical purposes the result is a set of truly random numbers.

A simple way to generate distribution specific random numbers is to set the cumulative distribution function equal to a unit, random number, and take the inverse. Consider the exponential distribution

$$F(x) = 1 - e^{-\lambda x} \tag{4.1}$$

By setting a random variable, uniformly distributed from zero to one, r, equal to the cumulative distribution function, $F(x)$, and inverting the function, an exponentially distributed random variable, x, with a failure rate of λ is created.

$$r = 1 - e^{-\lambda x}$$
$$1 - r = e^{-\lambda x}$$
$$\ln(1 - r) = -\lambda x \tag{4.2}$$
$$x = -\frac{\ln(1 - r)}{\lambda}$$

This expression can be further reduced; the term $1-r$ is also uniformly distributed from zero to one. The result is

$$x = -\frac{\ln r}{\lambda} \tag{4.3}$$

Table 4.1   Random number generators.[2]

Distribution	Probability Density Function	Random Number Generator
Uniform	$f(x) = \dfrac{1}{b-a}, \quad a \le x \le b$	$x = a + (b-a)r$
Exponential	$f(x) = \lambda e^{-\lambda x}, \quad 0 < x < \infty$	$x = -(1/\lambda) \ln r$
Normal	$f(x) = \dfrac{1}{\sigma\sqrt{2\pi}} \exp\left[-\dfrac{1}{2}\left(\dfrac{x-\mu}{\sigma}\right)^2\right],$ $-\infty < x < \infty$	$x_1 = \left[\sqrt{-2\ln r_1}\,\cos\left(2\pi r_2\right)\right]\sigma + \mu$
Lognormal	$f(x) = \dfrac{1}{\sigma x\sqrt{2\pi}} \exp\left[-\dfrac{1}{2}\left(\dfrac{\ln x-\mu}{\sigma}\right)^2\right],$ $x > 0$	$x_1 = \exp\left[\sqrt{-2\ln r_1}\,\cos\left(2\pi r_2\right)\right]\sigma + \mu$
Weibull	$f(x) = \dfrac{\beta x^{\beta-1}}{\theta^\beta} \exp\left(-\dfrac{x}{\theta}\right)^\beta, \quad x > 0$	$x = \theta(-\ln r)^{1/\beta}$

This is done easily in electronic spreadsheets by using distribution inverse functions. In Excel® the function =RAND() yields a unit uniform random number. To achieve a random number from a normal distribution use the function

$$= \mathrm{NORM.INV}(\mathrm{RAND}(), \mu, \sigma)$$

where μ and σ are the mean and standard deviation of the distribution.

The inverse functions typically include a numerical approximation that makes a simulation slow. A more efficient method is to use the closed form random generators given in Table 4.1.

In addition to the expressions from Table 4.1, many programming languages have functions to generate random numbers. The Boost library (Boost.org) provides many mathematical routines for C++ including random number generators.

4.3   Validation

After the desired random number, generator(s) has been constructed, the next step is to mathematically model the situation under study. After completing the model, it is important to validate the model. A valid model is a reasonable representation of the conceptual model being studied. The simulation model is validated by confirming each of the inputs, and ensuring the calculations in the simulation model are correct. A common procedure for validation is:

1. compare the mean and standard deviation of each of the inputs to their theoretical values,

---

[2]The random number generators given in this table are contained in a spreadsheet on the accompanying website. The spreadsheet is titled RandomNumberGenerators.xls and there is a tab in the spreadsheet for each distribution.

2. construct histograms for each of the inputs and compare the histogram shape to the expected distribution shape (symmetrical, skewed right, skewed left, etc.), and

3. randomly select a few rows from the simulation, and ensure the outputs are computed correctly.

Enough iterations should be included in the simulation to provide a steady-state solution. A steady-state solution is reached when the output of the simulation from one iteration to the next changes negligibly or the change is less than the allowable error. When calculating means and variances, 1000 iterations are often sufficient. If calculating process capability, many more iterations may be required. At least one million iterations are required to have a reasonable result when establishing that a design is capable of meeting specifications with a failure rate less than 10 parts per million.

Example 4.1[3]

Demonstrate the basic concepts of simulation by simulating a coin flip.

Solution

Using Excel® a coin flip can be simulated using the function below. The *Rand()* function returns a uniform random number between zero and one. The *If* function returns the first term after the comma if true and the second term after the comma if false. The *If* function makes a logical comparison to 0.5 because a coin should land on *Heads* 50% of the time.

$$= \text{IF(RAND()} < 0.5, \text{``Heads'', ``Tails''})$$

Text output such as *Heads* and *Tails* is not as useful as numerical output. If *Heads* is assigned a value of zero and *Tails* is assigned a value of one, the average of the iterations yields the percentage of tails. This example is simulated using Excel® with the following steps.

1. In cell A1 place the function =IF(RAND()<0.5,1,0).

2. In cell B1 place the function =AVERAGE(A$1:A1).

3. In cell A2 place the function =IF(RAND()<0.5,1,0).

4. In cell B2 place the function =AVERAGE(A$1:A2).

5. Copy the second row 1000 times.

Each row of the spreadsheet simulates the physical act of flipping a coin. Copying the row 1000 times simulates flipping an actual coin 1000 times. Column B can be used to determine when steady state is reached. Figure 4.2 displays a running average of the percentage of tails. Note the erratic behavior when the number of trials is low.

---

[3] This example is included on the accompanying website. The Excel® file is "Simulation_CoinFlip.xls," and the Sage® file is "SageCoinFlipSimulation.txt."

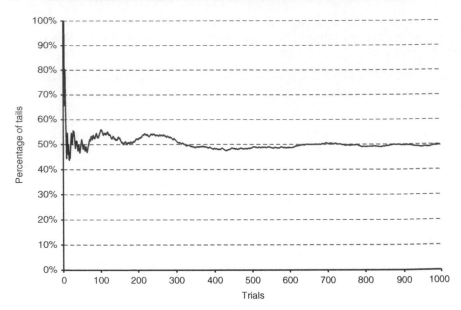

Figure 4.2   Simulation of a coin flip.

Example 4.2[4]

In Chapter 3, the simulation results of the power of a simple circuit were given. The power in this circuit is

$$P = \frac{V^2}{R}$$

The requirements for power are 72 ± 11. The nominal voltage is 12 with a standard deviation of 0.2, and the nominal resistance is 2 ohms with a standard deviation of 0.1. Since there is no reason that voltage or resistance would be skewed, a normal distribution is assumed for both parameters (of course this could be verified by taking some samples). Using Monte Carlo simulation, determine the percentage of circuits that do not meet the power specification.

Solution

This example is simulated using Excel® with the following steps.

1. In cell A1 place the formula (generate a normal random number for voltage).

$$= 0.2*((COS(2*PI()*RAND()))*(-2*LN(RAND()))^0.5) + 12$$

---

[4] This example is included on the accompanying website. The Excel® file is "CircuitPowerSimulation.xls," and the Sage® file is "SageCircuitPowerSimulation.txt."

2. In cell B1 place the formula (generate a normal random number for resistance).

$$= 0.1^*((COS(2^*PI()^*RAND()))^*(-2^*LN(RAND()))^{\wedge}0.5) + 2$$

3. In cell C1 place the formula (compute power from voltage and resistance).

$$= (A1^{\wedge}2)/B1$$

4. In cell D1 place the function (places a 1 in the cell if the power is less than the lower specification).

$$= IF(C1 < 61, 1, 0)$$

5. In cell E1 place the function (places a 1 in the cell if the power is greater than the upper specification).

$$= IF(C1 > 83, 1, 0)$$

6. Copy this row 1000 times.

Before using the results of any simulation, the simulation must be validated. In this case, it must be verified that voltage follows a normal distribution with a mean of 12 and a standard deviation of 0.2, and that resistance follows a normal distribution with a mean of 2 and a standard deviation of 0.1. This was done in Chapter 3, as histograms of the simulation results for voltage and resistance were created. The calculations can be validated by randomly selecting a few simulation iterations and ensuring the calculations are correct.

The average of column D1 estimates the percentage of circuits that fall below the lower specification. The average of column E1 estimates the percentage of circuits that exceed the upper specification. For this circuit approximately 0.2% of all circuits fall below the lower specification and approximately 1.1% of all circuits exceed the upper specification. The probability of failing the lower specification is not the same as the probability of failing the upper specification, because the distribution of power is not symmetrical even though the distribution of voltage and resistance is symmetrical. This is shown in Figure 4.3 as the deviation from the target (72 watts) is greater on the high side than it is on the low side.

From the simulation, the average power is 72.22 watts with a standard deviation of 4.34. When the output is skewed it is common for the actual output and the theoretical output to be unequal. For the circuit, the theoretical power is 72 ($12^2/2$). The difference between the actual output and the theoretical output is statistical bias caused by statistical variation in a nonlinear system as described in Chapter 3.

Observing histograms and counting out-of-specification occurrences yield a great deal of information concerning the behavior of a stochastic system. In some cases, it is useful to supplement this information by fitting a distribution to the simulation data. When variables are multiplied and divided the resulting distribution often follows a

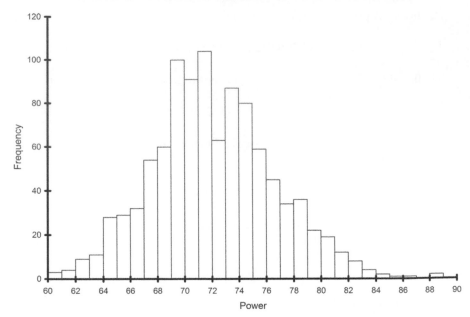

Figure 4.3   Histogram of simulated power.

lognormal distribution. The histogram shown in Figure 4.3 is right skewed which is consistent with the behavior of a lognormal distribution, but further evidence is needed. A lognormal probability plot is given in Figure 4.4. Since the plotted points follow a straight line with no irregular patterns, the lognormal distribution can be accepted as a reasonable statistical model for the power.

4.4   Stratified sampling

In some cases, Monte Carlo simulation is not feasible due to the time required to complete each iteration. For example, complex modeling of lubricants and metal fatigue may require several hours of computer time to achieve the desired outputs for a single iteration of a Monte Carlo simulation. Stratified sampling is a technique that can be used to dramatically reduce the number of iterations with only a slight loss in accuracy.

When using stratified sampling, values are systematically chosen across the distribution range rather than chosen randomly. After the values are chosen, they are placed in random order and combined randomly with other variables. At least 100 iterations are recommended, but as few as 10 iterations may be used. The accuracy of the simulation improves as the number of input variables increases as well as when the number of iterations is increased.

The procedure for performing a stratified sampling simulation is:

1. define the system to be simulated,

2. determine the variation behavior of the inputs to the system,

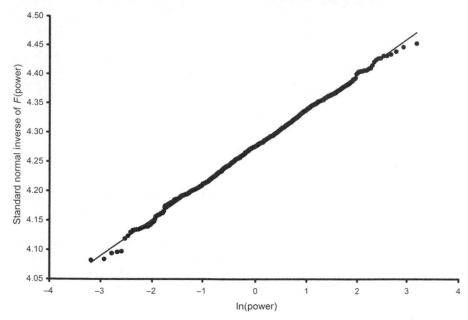

Figure 4.4   Lognormal probability plot for power.

3. determine percentiles for each of the n iterations in the simulation,

$$p_1 = \frac{1}{2n} \tag{4.4}$$

$$p_i = p_{i-1} + \frac{1}{n}, i > 1 \tag{4.5}$$

4. for each input to the system, determine stratified values representative of the input's variation by taking the inverse of the representative probability density function using the percentiles from step 3,

5. for each input to the system, place the stratified values from step 4 in random order,

6. compute the system output(s),

7. estimate the parameters of the appropriate statistical distribution, and

8. estimate the probability of not meeting specification limits.

Example 4.3[5]

Repeat Example 4.2 using stratified sampling with 20 iterations.

---

[5] This example is contained on the accompanying website. The file name is CircuitPowerSimulation-Stratified.xls.

Solution

The percentile for the first iteration is

$$p_1 = \frac{1}{(2)(20)} = 0.025$$

The percentile for the remaining iterations is

$$p_i = 0.025 + \frac{i-1}{20}$$

The percentiles along with the inverse of the normal probability density are given in Table 4.2.

Before power can be computed, the stratified values of voltage and resistance must be randomized. This step is required for a realistic output since the nonrandomized data match the smallest voltage to the smallest resistance, the largest voltage to the largest resistance, and the equivalent ordering for all values. Randomized voltage, resistance, and the resulting power are given in Table 4.3.

Table 4.2   Circuit power stratified sampling data.

Iteration	Percentile	Voltage	Resistance
1	0.025	11.608	1.804
2	0.075	11.712	1.856
3	0.125	11.770	1.885
4	0.175	11.813	1.907
5	0.225	11.849	1.924
6	0.275	11.880	1.940
7	0.325	11.909	1.955
8	0.375	11.936	1.968
9	0.425	11.962	1.981
10	0.475	11.987	1.994
11	0.525	12.013	2.006
12	0.575	12.038	2.019
13	0.625	12.064	2.032
14	0.675	12.091	2.045
15	0.725	12.120	2.060
16	0.775	12.151	2.076
17	0.825	12.187	2.093
18	0.875	12.230	2.115
19	0.925	12.288	2.144
20	0.975	12.392	2.196

Table 4.3   Randomized circuit power stratified sampling data.

Iteration	Voltage	Resistance	Power
1	11.813	1.885	74.033
2	11.880	1.955	72.211
3	12.038	2.196	65.988
4	12.091	2.019	72.408
5	11.987	1.924	74.670
6	11.849	1.981	70.869
7	11.712	2.115	64.856
8	11.936	2.144	66.454
9	12.013	1.994	72.378
10	11.909	2.060	68.857
11	12.151	2.006	73.594
12	12.120	2.032	72.290
13	11.608	1.968	68.464
14	12.392	1.940	79.146
15	12.288	2.076	72.749
16	12.064	2.045	71.152
17	11.962	1.804	79.320
18	11.770	2.093	66.173
19	12.187	1.907	77.901
20	12.230	1.856	80.588

The average of the power from Table 4.3 is 72.205 watts, and the power sample standard deviation is 4.608 watts. A lognormal probability plot, shown in Figure 4.5, verifies that the lognormal is an adequate statistical distribution for power as the points form a reasonably straight line with no discernable patterns.

The parameters of the lognormal distribution are estimated from the average and sample standard deviation of the stratified power data.

$$\hat{\sigma} = \sqrt{2 \ln[72.205] - \ln[(72.205)^2 - (4.608)^2]} = 0.064$$

$$\hat{\mu} = \ln[72.205] - \frac{(0.064)^2}{2} = 4.277$$

The percentage of circuits falling below the lower specification can be found with Excel® using the function[6]

$$P(\text{Power} < 61) = \text{LOGNORM.DIST}(61, 4.277, 0.064) = 0.5\%$$

---

[6] This function is for Excel® version 2010 and later. Earlier versions use the function LOGNORMDIST.

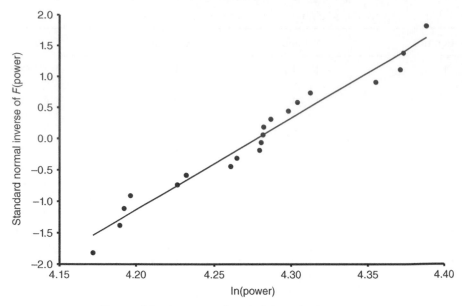

Figure 4.5   Lognormal probability plot for power.

The percentage of circuits exceeding the upper specification can be found with Excel® using the function

$$P(\text{Power} > 83) = 1 - \text{LOGNORM.DIST}(83, 4.277, 0.064) = 1.4\%$$

The estimated percentage of circuits not meeting specifications is reasonably close to the values found with Monte Carlo simulation despite only using 20 iterations.

4.5   Summary

Simulation is a powerful analytical tool that can be used to model virtually any system. For the above examples, 1000 iterations were used. The number of iterations used should be based on reaching a steady-state condition. Depending on the problem, more or less iterations may be needed.

Once simulation is mastered, a danger is that it is overused because of the difficulty involved with other mathematical models. In the following chapters, methods are shown to create mathematical models to estimate the mean and variance of systems consisting of a combination of random variables. These models can be verified and supplemented with simulation. These mathematical models also lend themselves to optimization, whereas optimization is much more difficult when using simulation.

Exercises

1 Use Monte Carlo simulation to estimate the standard deviation of the system below, given $\mu_x = 65$ and $\sigma_x = 2.8$.

$$y = 12.4 \left[\cos \left(\frac{x + 340}{2.7\pi} \right) \right], 0 < x < 20$$

2 The mean of x_1 is 8.2 and the standard deviation of x_1 is 0.2. The mean of x_2 is 4.7 and the standard deviation of x_2 is 0.5. Given $y = e^{(x_1/x_2)}$, use Monte Carlo simulation to estimate the standard deviation of y.

3 Use Monte Carlo simulation to estimate the standard deviation of the system below, given $\mu_V = 12$, $\sigma_V = 0.1$, $\mu_R = 3$, and $\sigma_R = 0.2$.

$$I = \frac{V}{R}$$

4 Given a lower specification of 3.7 amps and an upper specification of 4.3 amps, estimate the percentage of circuits from Exercise 4.3 not meeting specifications.

5 Given force is equal to the product of mass and acceleration, what is the mean and standard deviation of force, given $\mu_{mass} = 20$, $\sigma_{mass} = 0.4$, $\mu_{acceleration} = 5$, and $\sigma_{acceleration} = 0.2$.

6 The deflection at the free end of a cantilever beam with a load of P at the free end is

$$\delta = \frac{4Pl^3}{Ebh^3}$$

where l is the beam length, E is the modulus of elasticity, b is the beam width, and h is the beam height.

Determine the percentage beams that have a deflection at the free end greater than 170 µm given P(Newtons) has a Weibull distribution with parameters $\beta = 8.5$, $\theta = 500$, l(meters) is normally distributed with a mean of 15 and a standard deviation of 0.09, E(Pascals) is normally distributed with a mean of $2.0685(10^{11})$ and a standard deviation of $2(10^5)$, b(meters) is normally distributed with a mean of 0.5 and a standard deviation of 0.02, and h(meters) is normally distributed with a mean of 0.8 and a standard deviation of 0.025.

7 Repeat Exercise 4.3 using stratified sampling.

8 Repeat Exercise 4.4 using stratified sampling.

5

Modeling variation of complex systems

Before releasing a design, engineers must validate the design against all tolerances. Since a finished product is often a combination of many individual components, this task can be difficult. One approach to validation is to analyze worst-case tolerances. This task can be daunting as the number of combinations dramatically increases as the number of tolerances specified increases. For example, a simple design with 10 tolerances has 1024 possible combinations of high and low tolerance settings. A design with 20 tolerances has over one million combinations of high and low tolerance settings. In addition, as shown in Chapter 3, worst-case tolerance analysis is unrealistic. The likelihood of a part being produced with 20 tolerances all at the edge of the tolerance limit is highly unlikely.

Engineers need a method for modeling the variation of their design as a function of the variation of the inputs. This type of model not only allows validation, but also allows engineers to optimize the design against variation in the inputs and to identify the noncritical components for cost reduction efforts. Models for estimating system variance and expected value for simple systems were presented in Chapter 3. Mathematical complexity for these methods becomes overwhelming for all but the simplest engineering systems. Monte Carlo simulation, presented in Chapter 4, can be used to predict the variance and expected value for even the most complex systems, but the number of computations involved makes it difficult to use Monte Carlo simulation models with optimization routines. In addition, variance contributions for each input variable are not easily obtained from simulations. This chapter presents a technique for estimating the expected value and variance of the system, including variance contributions for each input variable.

Probabilistic Design for Optimization and Robustness for Engineers, First Edition.
Bryan Dodson, Patrick C. Hammett and René Klerx.
© 2014 John Wiley & Sons, Ltd. Published 2014 by John Wiley & Sons, Ltd.
Companion website: http://www.wiley.com/go/robustness_for_engineers

5.1   Approximating the mean, bias, and variance

Taylor series expansions of moment generating functions can be used to overcome the mathematical complexities of the methods presented in Chapter 3 to obtain estimates for the expected value and variance of functions of several variables. Given a function of several variables

$$y = g(x_1, x_2, \ldots, x_n) \tag{5.1}$$

a Taylor series expansion about the mean is

$$
\begin{aligned}
y \approx g[E(x_1), E(x_2), \ldots, E(x_n)] &+ \sum_{i=1}^{n} [x_i - E(x_i)] \frac{\partial g}{\partial x_i} \\
&+ \frac{1}{2} \sum_{i=1}^{n} \sum_{j=1}^{n} [x_i - E(x_i)][x_j - E(x_j)] \frac{\partial^2 g}{\partial x_i \partial x_j} + \cdots
\end{aligned}
\tag{5.2}
$$

When input variation is low or the function is not highly nonlinear, the first-order of the Taylor expansion can be used to obtain reasonable estimates for the mean and variance.

$$E(y) \approx g[E(x_1),\ E(x_2), \ldots, E(x_n)] \tag{5.3}$$

$$V(y) \approx \sum_{i=1}^{n} \sum_{j=1}^{n} \left(\frac{\partial g}{\partial x_i} \right) \left(\frac{\partial g}{\partial x_j} \right) \mathrm{Cov}\left(x_i,\ x_j \right) \tag{5.4}$$

If the input variables are independent, the variance estimate reduces to

$$V(y) \approx \sum_{i=1}^{n} \left[\left(\frac{\partial g}{\partial x_i} \right)^2 \sigma_{x_i}^2 \right] \tag{5.5}$$

When highly nonlinear systems or systems with excessive input variance are encountered, a second term from the Taylor series expansion may be needed to obtain estimates with an acceptable error level. If the input variables are independent, second-order Taylor series approximations for the mean and variance are:

$$E(y) \approx g[E(x_1),\ E(x_2), \ldots, E(x_n)] + \frac{1}{2} \sum_{i=1}^{n} \left[\left(\frac{\partial^2 g}{\partial x_i} \right) \sigma_{x_i}^2 \right] \tag{5.6}$$

$$V(y) \approx \sum_{i=1}^{n} \left[\left(\frac{\partial g}{\partial x_i} \right)^2 \sigma_{x_i}^2 \right] + \frac{1}{2} \sum_{i}^{n} \sum_{j}^{n} \left[\left(\frac{\partial^2 g}{\partial x_i \partial x_j} \right)^2 \sigma_{x_i}^2 \sigma_{x_j}^2 \right] \tag{5.7}$$

The first term of Equation 5.6 is often excluded resulting in the difference between the theoretical and actual values of the function, which is referred to as the bias.

Table 5.1   Circuit input data.

Parameter	Nominal	Standard Deviation	Distribution
Voltage	12	0.2	Normal
Resistance	2	0.1	Normal

The second-order Taylor series approximations improve accuracy at the cost of additional mathematical computations, however, far fewer additional computations are required for estimating the mean than the variance. Because of this, it is recommended to use the second-order approximation for the mean and the first-order approximation for the variance as a starting point, and introduce the second-order approximation for the variance if accuracy must be improved.

Example 5.1[1]

Consider the circuit simulated in Chapter 3 consisting of a single power supply and a single resistor. The power in this circuit is

$$P = \frac{V^2}{R} \tag{5.8}$$

Given the input values to the circuit in Table 5.1, estimate the average power and the standard deviation of the power.

Solution

To develop an equation for the expected value and standard deviation of power using Equations 5.5 and 5.6, the first and second derivatives of power with respect to voltage and resistance are needed.

$$\frac{dP}{dV} = \frac{2V}{R} \tag{5.9}$$

$$\frac{d^2P}{dV^2} = \frac{2}{R} \tag{5.10}$$

$$\frac{dP}{dR} = -\frac{V^2}{R^2} \tag{5.11}$$

$$\frac{d^2P}{dR^2} = \frac{2V^2}{R^3} \tag{5.12}$$

---

[1] Solutions to this example are contained on the accompanying website. The Excel® solution is named CircuitPowerModel.xls. The Sage® solution is named SageCircuitPower.txt.

	A	B	C	D	E	F	G	H
			Input				Output	
			Standard	First	Second	Output	Standard	
1	Parameter	Nominal	Deviation	Derivative	Derivative	Variance	Deviation	Bias
2	Voltage	12	0.2	12.00	1.00	5.76	2.40	0.02
3	Resistance	2	0.1	-36.00	36.00	12.96	3.60	0.18
4	Power (theoretical)	72.00						
5	Power (actual)	72.20						
6	Power (standard deviation)	4.33						

Figure 5.1   Circuit power calculations in Excel®.

The estimated power in the circuit is

$$E(P) \approx \frac{V^2}{R} + \frac{1}{2}\left[\left(\frac{2}{R}\right)(\sigma_V^2) + \left(\frac{2V^2}{R^3}\right)(\sigma_R^2)\right]$$

$$\approx \frac{12^2}{2} + \frac{1}{2}\left[\left(\frac{2}{2}\right)(0.2^2) + \left(\frac{(2)12^2}{2^3}\right)(0.1^2)\right] \approx 72.2$$

The estimated variance in the circuit is

$$V(P) \approx \left(\frac{2V}{R}\right)^2(\sigma_V^2) + \left(-\frac{V^2}{R^2}\right)^2(\sigma_R^2)$$

$$V(P) \approx \left(\frac{(2)12}{2}\right)^2(0.2^2) + \left(-\frac{12^2}{2^2}\right)^2(0.1^2) \approx 18.72$$

The estimated standard deviation of the circuit power is

$$\sigma_P \approx \sqrt{18.72} \approx 4.33$$

It is useful to arrange the above calculations in a tabular format. This allows the variation contribution of each of the input variables to be easily quantified. Figure 5.1 displays a tabular summary of the circuit power calculations from Excel®. It can be seen in column G of the spreadsheet that the variability of the resistance contributes more (3.6) to the power standard deviation (4.33) than the variability of the voltage (2.4). Note that the variance in column F sums to obtain the variance of power. The resulting standard deviation is the square root of the variance.

The formulas used to obtain the values in Figure 5.1 are shown in Figure 5.2.

Number Empire[2] can be used to facilitate calculating derivatives. The formula from cell B4 in the spreadsheet shown in Figure 5.2 can be pasted into the Number Empire tool, and the resulting derivative formula can be pasted into Excel®. Figure 5.3 shows the input screen for the Number Empire derivative tool. Note that "B3" is in the

---

[2] Number Empire (http://www.numberempire.com/) provides a free online derivative calculation tool that provides output in both symbolic form and a form that can be pasted directly into Excel®.

	A	B	C	D	E	F	G	H
	Parameter	Nominal	Input Standard Deviation	First Derivative	Second Derivative	Output Variance	Output Standard Deviation	Bias
1	Parameter	Nominal	Deviation	Derivative	Derivative	Variance	Deviation	Bias
2	Voltage	12	0.2	=2*B2/B3	=2/B3	=C2^2*D2^2	=F2^0.5	=0.5*E2*C2^2
3	Resistance	2	0.1	=-(B2^2)/(B3^2)	=2*B2^2/B3^3	=C3^2*D3^2	=F3^0.5	=0.5*E3*C3^2
4	Power (theoretical)	=B2^2/B3						
5	Power (actual)	=B4+SUM(H2:H3)						
6	Power (standard deviation)	=SUM(F2:F3)^0.5						

Figure 5.2   Circuit power formulas in Excel®.

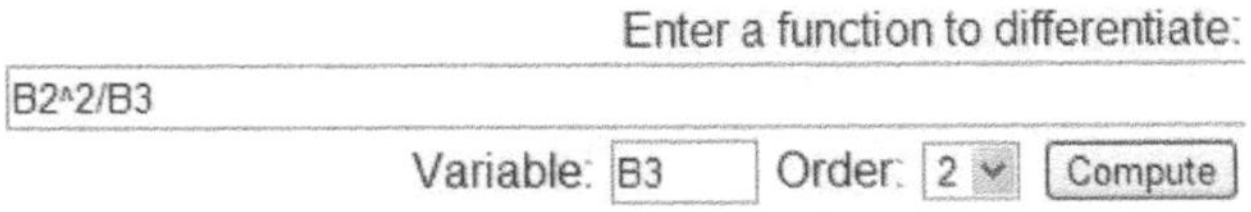

Enter a function to differentiate:

B2^2/B3

Variable: B3   Order: 2 ▾   [Compute]

Figure 5.3   Number Empire derivative calculator input.

2*B2^2/B3^3

$$\frac{\partial^2 f}{\partial B_3{}^2} = \frac{2\,B_2{}^2}{B_3{}^3}$$

Figure 5.4   Number Empire derivative calculator output.

variable box. In the Excel® spreadsheet, cell B3 represents resistance, so the figure shows the input to obtain the second derivative with respect to resistance. Beware that Number Empire is case sensitive.

Figure 5.4 shows the output of the Number Empire derivative calculator. The second line is in a format that can be pasted directly into Excel®. To avoid copying the formatting, the formula can be pasted directly into the formula bar or the paste special command may be used.

Using the above formulas to compute a mean and standard deviation has advantages over simulation in that the calculations are faster and easier to use with optimization routines, however, there is no information related to the shape of the distribution. It is recommended to use the above calculations in conjunction with simulation as validation, and to aid in determining the statistical distribution of the output. From the simulation in Chapter 4, the power in the circuit was estimated to be 72.22 with a standard deviation of 4.34. The simulation results are nearly identical to the values calculated above.

With a power specification of 72 ± 11 simulation predicted a failure rate of 1.3% with 1.1% of the circuits falling above the upper specification limit. The failure rate estimated with simulation was done nonparametrically by counting the failures. Failure rates can also be found with the results of Equations 5.5 and 5.6 by assuming a distribution and using the output mean and standard deviation to determine the parameters of the assumed distribution.

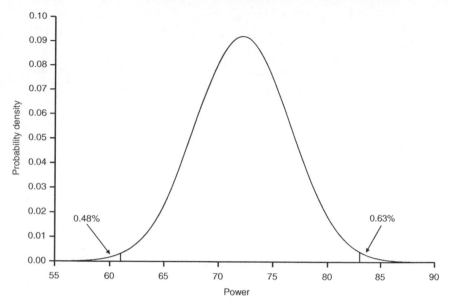

Figure 5.5   Normal probability density function for circuit power.

If a normal distribution is assumed, then the total failure rate is estimated at 1.1% with 0.63% of the circuits falling above the upper specification limit. This is reasonably close to the simulated failure rate of 1.3%. Unless the failure rate is exceedingly high, the normal distribution usually provides an adequate approximation. Figure 5.5 shows the failure rate graphically, and the failure rates can be found with the following Excel® functions.[3]

$$P(\text{Power} < 61) = \text{NORM.DIST}(61, 72.2, 4.33, 1)$$
$$P(\text{Power} > 83) = 1 - \text{NORM.DIST}(83, 72.2, 4.33, 1)$$

5.2   Estimating the parameters of non-normal distributions

Although the normal distribution provides reasonable estimates in many cases, it is not suitable in all situations. When data are skewed, the lognormal or Weibull distributions will provide more accurate probability estimations. The parameters for the lognormal or Weibull distribution can be determined by using the estimated mean and variance and matching these with theoretical values as shown in Chapter 2.

---

[3] The function NORM.DIST is only available in version 2013 and later. Use the function NORMDIST for earlier versions.

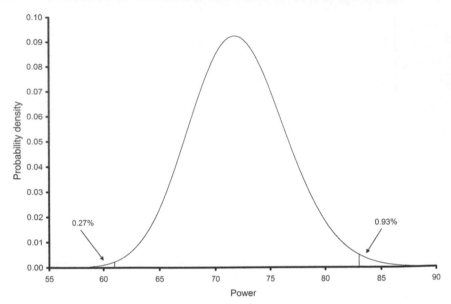

Figure 5.6   Lognormal probability density function for circuit power.

Example 5.2[4]

When variables are multiplied and divided, the result is often lognormally distributed. The power in a circuit results from a formula with multiplication and division, and it is reasonable to assume a lognormal distribution for power. Estimate the parameters of the lognormal distribution for the power of the circuit described in Example 5.1.

Solution

From Example 5.1, the expected power is estimated to be 72.2 amps with an estimated standard deviation of 4.33. The parameters of the lognormal distribution are:

$$\sigma = \sqrt{2 \ln(72.2) - \ln(72.2^2 - 4.33^2)} = 0.06$$
$$\mu = \ln(72.2) - \frac{0.06^2}{2} = 4.28$$

If a lognormal distribution is assumed, then the total failure rate is estimated at 1.2% with 0.93% of the circuits falling above the upper specification limit. This is reasonably close to the failure rate of 1.3% found with simulation, and the split between the high and low side of the specification is closer to the simulated values than the estimates made using the normal distribution. Figure 5.6 shows the failure rate

---

[4] This example is contained on the accompanying website. The Excel® solution is named Circuit-PowerModelLognormal.xls. The Sage® solution is named LognormalAndWeibullParameters.txt.

graphically. These failure rates can be found with the following Excel® functions.[5]

$$P(\text{power} < 61) = \text{LOGNORM.DIST}(61, 4.28, 0.06, 1)$$
$$P(\text{power} > 83) = 1 - \text{LOGNORM.DIST}(83, 4.28, 0.06, 1)$$

Example 5.3[6]

Find the parameters of the Weibull distribution for the power of the circuit in Example 5.1, and estimate the probability of not meeting the specification limits.

Solution

The parameters of the Weibull distribution for the power of the circuit in Example 5.1 can be estimated by solving the following equation for β.

$$4.33^2 = \left[\frac{72.2}{\Gamma\left(1 + \frac{1}{\beta}\right)} \right]^2 \left[\Gamma\left(1 + \frac{2}{\beta}\right) - \Gamma^2\left(1 + \frac{1}{\beta}\right) \right] \Rightarrow \beta = 20.7$$

The scale parameter of the Weibull distribution is estimated to be

$$\theta = \frac{72.2}{\Gamma\left(1 + \frac{1}{20.7}\right)} = 74.1$$

The probability of falling below the lower specification limit is

$$P(\text{power} < 61) = 1 - e^{-\left(\frac{61}{74.1}\right)^{20.7}} = 0.0176$$

The probability of falling above the upper specification limit is

$$P(\text{power} > 83) = e^{-\left(\frac{83}{74.1}\right)^{20.7}} \approx 0$$

The Weibull distribution does not match the simulation results very well, especially when computing the probability of power exceeding the upper specification.

---

[5] The function LOGNORM.DIST is only available in version 2013 and later. Use the function LOG-NORMDIST for earlier versions.

[6] This example is contained on the accompanying website. The Excel® solution is named Circuit-PowerRobustnessWeibull.xls. The Sage® solution is named LognormalAndWeibullParameters.txt.

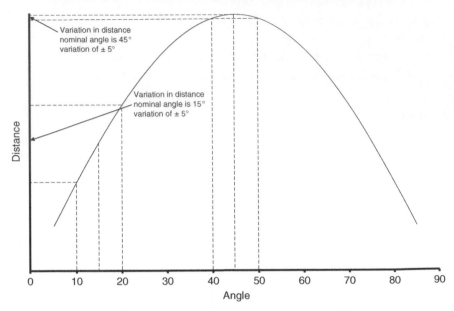

Figure 5.7   Poor variance estimation example.

Failure rates for the Weibull distribution can be found with the following Excel®
functions.[7]

$$P(\text{power} < 61) = \text{WEIBULL.DIST}(61, 20.7, 74.1, 1)$$
$$P(\text{power} > 83) = 1 - \text{WEIBULL.DIST}(83, 20.7, 74.1, 1)$$

5.3   Limitations of first-order Taylor series approximation for variance

Equation 5.5 uses the rate of change with respect to each input and the magnitude
of the change of each input to estimate the output variance. This equation assumes
the variance impact of higher-order effects and interaction between inputs is zero.
Consider the distance a projectile travels in a vacuum as a function of angle. If
gravity and velocity are held constant, the distance the projectile travels is shown
in Figure 5.7. Since the derivative is equal to zero when the angle is 45°, Equation
5.5 estimates zero variation in the projectile distance regardless of the magnitude of
angle variation.

---

[7] The function WEIBULL.DIST only works in version 2013 and later. Use the function WEIBULL
for earlier versions.

Table 5.2   Projectile input data.

Parameter	Nominal	Standard Deviation	Distribution
Angle	45° ($\pi/4$ radians)	5° ($5\pi/180$ radians)	Normal
Initial velocity	250 feet/s	4	Normal

A more accurate estimate of system variance is obtained by using the second-order of the Taylor series expansion.

Example 5.4[8]

Consider the distance a projectile travels as discussed above. The distance the projectile travels is

$$D = \frac{v^2 \left[\sin\left(2\theta\right)\right]}{g} \tag{5.13}$$

where v is the initial velocity, θ is the angle, and g is the gravitational acceleration.

Given the input values in Table 5.2, estimate the mean and the variance of the projectile using the first-order of the Taylor series expansion, and compare the results to the estimates using the first and second terms of the Taylor series expansion.

Solution

The first and second derivatives of distance with respect to initial velocity and angle are:

$$\frac{\mathrm{d}D}{\mathrm{d}\theta} = \frac{2v^2 \cos\left(2\theta\right)}{g} \tag{5.14}$$

$$\frac{\mathrm{d}^2 D}{\mathrm{d}\theta^2} = -\frac{4v^2 \sin\left(2\theta\right)}{g} \tag{5.15}$$

$$\frac{\mathrm{d}D}{\mathrm{d}v} = \frac{2v \sin\left(2\theta\right)}{g} \tag{5.16}$$

$$\frac{\mathrm{d}^2 D}{\mathrm{d}v^2} = \frac{2 \sin\left(2\theta\right)}{g} \tag{5.17}$$

---

[8] This example is contained on the accompanying website. The Excel® solution is named ProjectileDistanceModel.xls. The Sage® solution with simple variance estimation is named SageProjectileModel.txt. The Sage® solution estimating variance with higher-order effects is named SageProjectileModelHigherOrder.txt.

Table 5.3 Projectile distance calculations (first-order Taylor series).

Parameter	Nominal	Input Standard Deviation	First Derivative	Second Derivative	Output Variance	Output Standard Deviation	Bias
Angle (radians)	0.7854	0.0873	0.00	−7764.0	0.00	0.00	−29.6
Velocity	250	4	15.53	0.06	3858	62.1	0.50
Distance (theoretical)	1941.0						
Distance (actual)	1911.9						
Distance (standard deviation)	62.1						

The estimated distance the projectile will travel (in feet) is

$$E(D) \approx \frac{v^2 [\sin(2\theta)]}{g} + \frac{1}{2} \left[\left(-\frac{4v^2 \sin(2\theta)}{g} \right) (\sigma_\theta^2) + \left(\frac{2\sin(2\theta)}{g} \right) (\sigma_v^2) \right]$$

$$\approx \frac{250^2 [\sin(2(\pi/4))]}{32.2}$$

$$+ \frac{1}{2} \left[\left(-\frac{4(250^2) \sin(2(\pi/4))}{32.2} \right) \left(5 \left(\frac{\pi}{180} \right) \right)^2 + \left(\frac{2\sin(2(\pi/4))}{32.2} \right) (4^2) \right]$$

$$\approx 1911.9$$

The estimated variance of the projectile distance (in feet squared) is

$$V(D) \approx \left(\frac{2v^2 \cos(2\theta)}{g} \right)^2 (\sigma_\theta^2) + \left(\frac{2v\sin(2\theta)}{g} \right)^2 (\sigma_v^2)$$

$$V(D) \approx \left(\frac{2(250)^2 \cos(2(\pi/4))}{32.2} \right)^2 \left(5 \left(\frac{\pi}{180} \right) \right)^2 + \left(\frac{2(250)\sin(2(\pi/4))}{32.2} \right)^2 (4^2)$$

$$\approx 3858$$

The standard deviation of the projectile distance is

$$\sigma_D \approx \sqrt{3858} \approx 62.1 \text{ feet}$$

Table 5.3 summarizes the above calculations. The *Output Variance* column shows that angle variability contributes nothing to the distance variability. This occurs when the first derivative with respect to any variable is equal to zero. This is a limitation of the estimation method used which causes the output variance to be underestimated.

Equation 5.7 can be used to provide a more precise estimate of the projectile distance variance by including higher-order effects and variable interaction effects. In addition to the derivatives previously calculated, this equation requires the second derivative of distance with respect to angle and velocity.

$$\frac{d^2 D}{d\theta dv} = \frac{4v \cos(2\theta)}{g} \tag{5.18}$$

The estimated variance is

$$V(D) \approx \left(\frac{2v^2 \cos(2\theta)}{g} \right)^2 \sigma_\theta^2 + \left(\frac{2v\sin(2\theta)}{g} \right)^2 \sigma_v^2$$

$$+ \frac{1}{2} \left(-\frac{4v^2 \sin(2\theta)}{g} \right)^2 \sigma_\theta^4 + \frac{1}{2} \left(\frac{2\sin(2\theta)}{g} \right)^2 \sigma_v^4 + \frac{1}{2} \left(\frac{4v \cos(2\theta)}{g} \right) \sigma_\theta^2 \sigma_v^2$$

$$
= \left(\frac{2(250^2) \cos \left(2\frac{\pi}{4} \right)}{32.2} \right)^2 \left(5\frac{\pi}{180} \right)^2 + \left(\frac{2(250) \sin \left(2\frac{\pi}{4} \right)}{32.2} \right)^2 4^2
$$

$$
+ \frac{1}{2} \left(-\frac{4(250^2) \sin \left(2\frac{\pi}{4} \right)}{32.2} \right)^2 \left(5\frac{\pi}{180} \right)^4 + \frac{1}{2} \left(\frac{2 \sin \left(2\frac{\pi}{4} \right)}{32.2} \right)^2 4^4
$$

$$
+ \frac{1}{2} \left(\frac{4(250) \cos \left(2\frac{\pi}{4} \right)}{32.2} \right) \left(5\frac{\pi}{180} \right)^2 (4^2) \approx 5606
$$

The standard deviation of the projectile distance is

$$
\sigma_D \approx \sqrt{5606} \approx 74.9 \text{ feet}
$$

The above calculations are summarized in Table 5.4.

Monte Carlo simulation with 50 000 trials produces an average distance of 1911.9 feet with a standard deviation of 74.9 feet.

This example is exaggerated to demonstrate the differences in the methods for approximating system variance. With a nominal angle of 45° and a standard deviation of 5°, the angle will vary from 30° to 60°. This is obviously beyond any expectation. With a more realistic angle variation, there is only a slight difference between estimated variance when using only the first-order Taylor series and using second-order Taylor series. Figure 5.8 shows the difference between the two methods as a function of angle standard deviation. As seen in Figure 5.8, the error is relatively low using the first-order approximation unless the input standard deviation is exceptionally high. In general, this is true unless the function under consideration has a very steep peak.

Including the second-order term when approximating variance can become quite cumbersome. The number of required terms is equal to twice the number of parameters under consideration plus all two-way combinations of the input parameters.

$$
n_T = 2n_p + \binom{n_p}{2} \tag{5.19}
$$

where n_T is the number of terms required for the variance estimation and n_p is the number of input parameters.

The number of terms required to estimate the system variance using both first- and second-order terms is shown in Figure 5.9.

It is recommended to use the first-order of the Taylor series to approximate variance and confirm the results with simulation. Unless the error is unacceptable, there is no need to include the second-order term.

Table 5.4 Projectile distance calculations including the second-order of the Taylor series expansion for variance estimation.

Parameter	Nominal	Input Standard Deviation	First Derivative	Second Derivative	Output Variance	Output Standard Deviation	Bias
Angle (radians)	0.7854	0.0873	0.0	−7764.0	0.0	0.0	−29.6
Velocity	250	4	15.53	0.1	3857.9	62.1	0.5
Angle - Angle				−7764.0	1747.9	41.8	0.0
Velocity - Velocity				0.1	0.5	0.7	0.0
Angle - Velocity				0.0	0.0	0.0	0
Distance (theoretical)	1941.0						
Distance (actual)	1911.9						
Distance (standard deviation)	74.9						

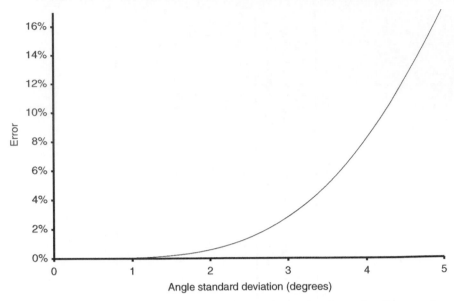

Figure 5.8   Error of the first-order variance estimate for the projectile example.

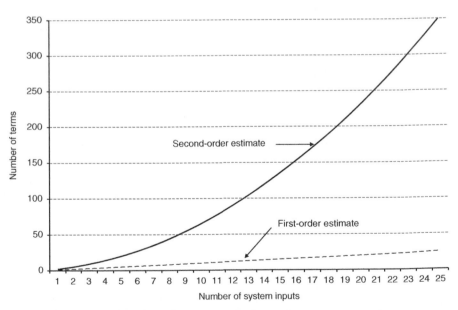

Figure 5.9   Number of terms required to estimate system variance.

When estimating the mean, using the second-order Taylor series approximation is not mathematically tedious. Although adding the second-order term does not greatly improve estimation accuracy, it is recommended to use the second-order term because the computational and modeling costs are low.

5.4   Effect of non-normal input distributions

The examples thus far have assumed a symmetrical distribution for the input parameters. Parameters such as roundness, surface finish, flatness, and many others have a skewed distribution. Nonsymmetrical distributions for input parameters have a small effect on the ability to predict the output standard deviation, and this error shrinks as the number of parameters in the model increases.

Example 5.5[9]

Consider the power in a circuit from Example 5.1. Instead of normal distributions for voltage and resistance, assume both voltage and resistance have an exponential distribution with the same means and standard deviations as Example 5.1. The exponential distribution is used in the example because it is drastically nonsymmetrical and is a near worst-case example. Illustrate the impact of nonsymmetrical distributions for input parameters by estimating the mean and standard deviation for power using Monte Carlo simulation and comparing the results to those found using normal distributions for voltage and resistance.

Solution

Figures 5.10 and 5.11 display the distributions of the voltage and resistance for a simulation of the power in 10 000 circuits, and Figure 5.12 gives a histogram for the resulting power. The mean and standard deviation for the voltage are identical to the values used in Example 5.1. Table 5.5 shows simulation results for circuit power using normal and exponentially distributed input parameters. Despite the tremendous difference in the distribution shape of the input parameters, the power results are close. By the power histograms created with normal and exponential inputs, the effect of the exponential inputs can be seen in the distribution shape, but the out-of-specification estimates are not great enough to have a meaningful impact on the optimization results.

This is a near worst-case for non-normal inputs. The exponential distribution is highly skewed, and there are only two inputs used to compute the output. As the number of inputs increases, the non-normal impact is reduced.

---

[9] This example is contained on the accompanying website. The Excel® solution is named Circuit-PowerModelExponentialInput.xls.

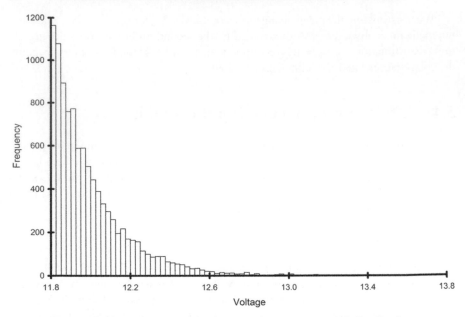

Figure 5.10   Histogram of voltage with an exponential distribution.

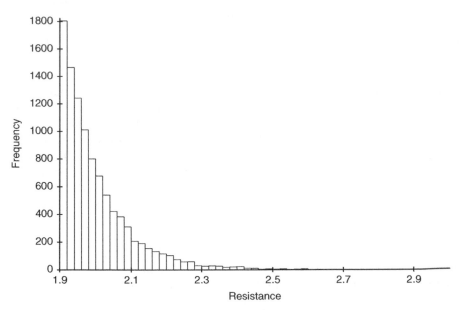

Figure 5.11   Histogram of resistance with an exponential distribution.

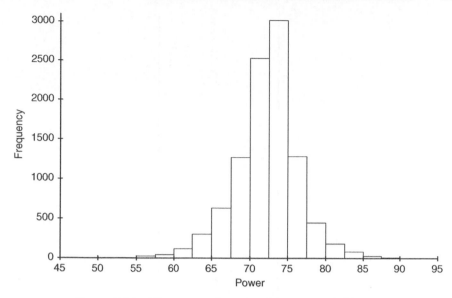

Figure 5.12   Histogram of power with exponential inputs.

5.5   Nonconstant input standard deviation

In the circuit and projectile examples above, the input standard deviations are constant; the standard deviations do not change when the nominal values change. A constant standard deviation is not required for the output standard deviation to be modeled and optimized (optimization is discussed in later chapters). The input standard deviation can be a percentage of the nominal, or a more complicated expression. In addition, nominal values of the input variables may be dependent on the nominal value or standard deviation of other input variables.

5.6   Summary

The mean, bias, and variance of a system can be estimated using the derivatives of the system function and the variance of the inputs. This method has advantages over simulation in that the equations are easier to manipulate mathematically when optimizing and when working with multiple outputs. Both normal and non-normal distributions may be used to estimate the mean, bias, and variance. A more accurate, higher-order method is also available, but the added accuracy is often small, and the higher-order method can become quite cumbersome.

The following chapters will extend these methods to optimize process capability and minimize system cost. Simple, single output systems as well as multiple output systems will be considered. In cases where equations do not exist, efficient experimentation methods are presented to empirically derive equations.

Table 5.5 Normal and exponential input comparison.

	Average of Simulated Values	Standard Deviation of Simulated Values	Power Less than Lower Specification	Power Greater than Upper Specification	Total Not Meeting Specification
Normal input distribution	72.22	4.34	0.47%	0.63%	1.10%
Exponential input distribution	72.12	4.06	1.22%	0.60%	1.82%

Exercises

1 The Taylor tool life model is used to calculate the length of time a cutting tool will last before a given amount of wear will occur. Tool life is

$$T = \left(C^{\frac{1}{n}} \right) \left(V^{-\frac{1}{n}} \right) \left(f^{-\frac{a}{n}} \right) \left(d^{-\frac{b}{n}} \right)$$

where V is the cutting velocity, f is the tool feed rate, d is the depth of cut, and C, n, a, and b are determined experimentally for each type of cutting tool.

A manufacturing company uses high speed steel (HSS) for its machining operations. From experimental data, the HSS tool life model constants are $C = 804$ $n = 0.17$, $a = 0.77$, and $b = 0.37$. The specification for tool life in the company are $21.75 + 4$. The nominal cutting velocity is 2 with a standard deviation of 0.05, the nominal value for tool feed is 0.3 with a standard deviation of 0.002, and the mean and standard deviation for depth of cut is 0.5 and 0.001, respectively. The input parameters follow a normal distribution. Assuming that the tool life follows a normal distribution, answer the following:

(a) Estimate the tool life using the second-order Taylor series approximation, and estimate the tool life standard deviation using the first-order Taylor series approximation.

(b) Assuming a normal distribution, estimate the percentage of tools below the specification limit for tool life and above the specification limit for tool life.

(c) Compare the answers from (a) and (b) with the Monte Carlo simulation results.

2 The equation which governs the power generated by a turbine is

$$P = 0.5 D A V^3$$

where D is the air density, A is the rotor swept area, and V is the wind velocity.

The data collected for the past five years from various locations where the turbine is installed show that the wind velocity and air density follow a normal distribution. The variation in rotor swept area is due to the process capability of the manufacturer and follows a normal distribution with a mean of 500 and standard deviation of 2. The mean air density is 1.2 with a standard deviation of 0.1, and the mean wind velocity is 20 with a standard deviation of 2. Industry standards dictate that a turbine having a rotor swept area as given above should generate a power output of $2.4(10^6) \pm 4(10^5)$.

(a) What percentage of turbines generates power below the minimum acceptable limit?

(b) What is the bias between the theoretical and actual value of power generated?

(c) Determine the parameters of the lognormal distribution for power using the estimated mean and standard deviation from (a).

(d) Using lognormal distribution, determine the percentage of turbines that generate power below the specification limit and above the specification limit.

(e) Compare the results from (b) and (d). Does the normal distribution provide reasonable estimates for percentage out-of-specification?

3 The amount of heat dissipated by a convective heat exchanger is governed by the equation

$$Q = \frac{(\Delta T)(A)}{C}$$

where ΔT is the temperature difference between the outside air and heat exchanger, A is the cross-sectional area of the heat exchanger, and C is the reciprocal of convective heat transfer coefficient.

The behavior of these variables is given below.

Variable	Mean	Standard Deviation
ΔT	20	0.9
A	100	0.6
C	0.5	0.1

(a) Assuming that C is normally distributed, the temperature difference and cross-sectional area are exponentially distributed, create a Monte Carlo simulation using 10 000 trails to estimate the heat dissipated and the standard deviation for the heat dissipated.

(b) Repeat (a) assuming a normal distribution for all input variables.

(c) Compare the results from (a) and (b). What can you conclude about the effect of nonsymmetrical distributions for input parameters on the ability to predict the output standard deviation?

4 A vending machine dispenses food packets by applying a force on them. If the force is too low the packet would get stuck and too high a force would crush the contents of the packet. Hence, the designers of the vending machine want to know the variation of the force, given the variation of the inputs, to tightly control the force within acceptance limits. The function for the force is

$$F = W^2 \sin(2A)$$

where W is a coefficient which depends on the packet size and weight and A is the angle at which the packet is placed.

The mean and standard deviation of the angle is 45° and 3°, respectively. The size coefficient has a nominal value of 54 and a standard deviation of 1.

(a) Determine the variance of the force using the first-order approximation.

(b) Determine the variance of the force using the second-order approximation.

(c) Compare the variance from (a) and (b). Had the designer used the first-order approximation, by how much would the variance have been underestimated?

6

Desirability

6.1 Introduction

In nearly all product design decision making, designers and managers must consider multiple requirements. Sometimes optimizing these requirements may be done independently, but often they involve trade-offs with other requirements. Consider the case of designing the portable 3D scanning equipment for measuring solid objects shown in Figure 6.1. Customers may desire a large field of vision to reduce the time for data capture. Customers may also want to reduce the size and weight of the equipment and maintain flexibility and portability for measuring objects of varying sizes in different environments.

We use this simple example to highlight challenges with multi-response optimization. For this scanning system, to improve the requirement metric, *field of vision*, the focal distance between cameras may need to be increased. This, in turn, may require increasing the length of the scanning equipment. If the scanner is too long, this reduces its portability and flexibility for measuring different objects in different usage environments.

To further complicate the design process, some potential customers may place greater importance on increasing field of vision (e.g., an aircraft manufacturer), whereas others may place a greater importance on portability (e.g., a headlamp manufacturer). As such, the designer may need to sub-optimize the performance for each application in an effort to appeal to multiple customer types.

Balancing requirements for multiple response outputs is also complicated by several other factors. For instance, requirements often have different units of measure. Field of vision may be measured by area versus the equipment weight which may be measured in kilograms. In addition, some requirements, such as portability, are difficult to quantify. And, perhaps most importantly, design decisions must consider costs

Probabilistic Design for Optimization and Robustness for Engineers, First Edition.
Bryan Dodson, Patrick C. Hammett and René Klerx.
© 2014 John Wiley & Sons, Ltd. Published 2014 by John Wiley & Sons, Ltd.
Companion website: http://www.wiley.com/go/robustness_for_engineers

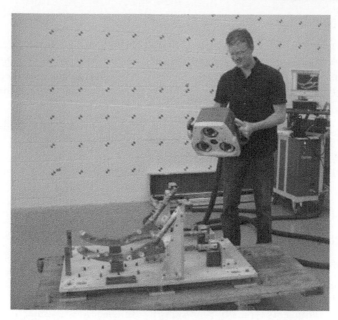

Figure 6.1   Measuring an object with 3D scanning system.

associated with maximizing performance, usability, and appeal for one requirement relative to the target cost of the overall product.

In this chapter, a multi-response optimization methodology to address these challenges is presented. This approach is based on a desirability function, originally developed by Harrington (1965), and later modified in Derringer & Suich (1980). This desirability function may be used to optimize a single requirement or to establish an overall total desirability index. Here, the desirability of multiple requirements is aggregated to provide an overall measure for a design solution.

In presenting this approach, we first discuss different types of requirements, criteria, and optimization conditions which may be summarized using a design scorecard (Yang & El-Haik, 2003). Next, we discuss estimating a desirability index for a single requirement using both a linear and nonlinear objective function. We then extend the case of single requirement desirability to the use of a total desirability index for multiple requirements and hence multi-response optimization. We conclude the chapter by exploring how to account for variation and robustness in assessing overall desirability of a design.

6.2   Requirements and scorecards

We begin with a discussion of different types of requirements and how to establish a design scorecard that summarizes performance expectations.

6.2.1   Types of requirements

In establishing requirements, we should first recognize that different types exist. We have requirements derived directly from the end customers buying or using a product. Typically, producers convert these customer requirements into technical requirements for product designers. These requirements may be expressed as functional performance metrics, such as measurement acquisition time, or as specific characteristics or features of the design solution (e.g., scanning system includes a calibration bar). In engineering and system modeling terminology, these requirements often are referred to as system outputs.

To further classify requirements, we may assign them into one of the following categories:

- end-customer requirements,

- functional requirements,

- design requirements, and

- process requirements.

Customer requirements typically represent the dimensionless, subjective wants and needs of external customers. For these requirements, organizations often measure preferences or customer satisfaction using surveys or other voice of the customer data collection techniques (Naumann & Hoisington, 2001). If customers are well educated about product choices and key performance metrics, companies may be able to use more quantitative measures to express customer requirements.

From customer requirements, designers often convert them into more quantifiable metrics known as *functional requirements*. These functional requirements may be used to measure the functional performance of a product, its usability[1], or design constraints. Figure 6.2 summarizes these functional requirements by categories and provides sample metrics for each. In developing functional requirements, we typically seek quantitative measures that may be evaluated relative to numerical goals or targets.

In addition, functional requirements, where possible, should be design neutral metrics to maximize the solution space for the final design. For example, an automotive company could avoid using the metric *lumens projected at 100 meters* for headlights, as this restricts the design solution to some type of lighting. In contrast, a metric such as *the ability to recognize a 3 m² object at 100 meters* allows more freedom for the designer.

Functional requirements flow down into *design requirements*. These requirements should be specific to a particular design solution. Here, one may identify metrics such as *lumens projected at 100 meters*. After flowing down to the lowest design level, these design requirements flow to the processes used to create them, and are expressed as key process input requirements. These may be referred to as *process requirements* or process parameters.

---

[1] Adapted from Nielsen (1993), Usability Engineering, Morgan Kaufman.

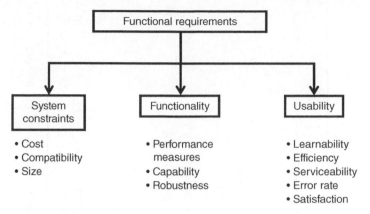

Figure 6.2   Types of functional requirements.

While the categorization of requirements may depend upon positioning within a product's value chain (component engineer versus system engineer perspective), the approach toward optimization may be viewed generically. One approach to displaying a set of requirements using any of the above categorizations is through a design scorecard.

6.2.2   Design scorecard

Design scorecards are used to describe a set of requirements and their desired performance goals. Scorecards are typically established at the initiation of a new product design project. Of course, some requirements and their target values may change or evolve during the product development life cycle. While one hopes to minimize changes to requirements through more effective market research and concept development activities, some change often is inevitable as new customer information becomes available. Thus, scorecards often represent a living document which must be regularly updated throughout the product life cycle.

In establishing a scorecard for requirements, designers need to identify the following:

- the response metric (preferably a quantitative measure),

- the target value for the response metric,

- acceptance limits for the response metric, and

- the direction of improvement (larger-the-better, smaller-the-better, nominal-is-best).

Advanced scorecards may include additional factors such as requirement categories, importance, goals for allowable variation, as well as other product metrics.

Table 6.1 provides an example of a design scorecard for a 3D scanning system. This scorecard is organized into different categories and includes importance ratings

Table 6.1 Advanced design scorecard for 3D scanning system grouped by categories.[2]

Requirement	Importance Rating	Direction of Improvement	Acceptance Criteria	Actual
Optical performance				
Field of view (cm^2)	4.5	Larger-the-better	$L = 900$; $T = 2600$	2500
Working distance (mm)	4.0	Nominal-the-best	780 ± 30	780
Image acquisition (μs)	4.5	Nominal-the-best	$T = 1 \pm 0.25$	1.1
Portability and flexibility				
Optical head weight (kg)	4.5	Smaller-the-better	$T = 9$; Upper $= 12$	10
Optical head size (cm^3)	4.0	Nominal-the-best	$40\,000 \pm 500$	40 200
Cord length (m)	3.0	Nominal-the-best	10 ± 1	10
Stress test reliability				
Mean-time-to-failure (hours)	4.0	Larger-the-better	$L = 375$; $T = 500$	395
Cost				
#Parts—Bill of material	3.5	Smaller-the-better	$T = 250$ K; Upper $= 400$	270
System cost ($)	4.0	Smaller-the-better	$T = 200$ K; Upper $= 300$ K	240

[2]The values used in this example are for instructional purposes only and not intended to represent product design decision trade-offs for a particular type of 3D scanning equipment.

for each requirement, acceptance criteria, and a desired direction of improvement. The importance ratings are based on survey data using a classic five-point scale (1 being least important and 5 being most important).

A critical step in multi-response optimization is to identify an appropriate target value, acceptance limits, and direction of improvement for each requirement. The target value represents the nominal or desired value to achieve. For instance, a vehicle may be assigned a target value of 40 miles per gallon at highway speed. The actual average miles per gallon achieved by a design solution will likely deviate from this target if other performance metrics represent real potential trade-offs. Thus, we also need to identify acceptance limits for optimization. These are critical because designers must be able to accept some deviations from initial design target values, particularly when cost implications are considered.

The term acceptance limits is preferred over specification limits. Acceptance limits imply design solution boundaries which may be deemed acceptable to a producer, whereas specification limits are used after a design is released to monitor ongoing quality or performance. The distinction between acceptance limits and specification limits also is important if a product has no specifications for a given requirement. In this case, one needs to define acceptance limits for desirability score calculations. An example of the lack of specification limits is the battery life of a cell phone. There is no fixed lower limit for batter life that renders a cell as unsalable.

The identification of acceptance limits is affected by the desired direction of improvement for each requirement metric. Directions of improvement for optimization are typically categorized into the following types with required acceptance limits:

- larger-is-better (target and lower limit),
- smaller-is-better (target and upper limit),
- nominal-is-best (target, lower limit, and upper limit), and
- nominal-is-best within a robust range around a target (target range, lower limit, and upper limit).

Figure 6.3 illustrates each of these conditions with their required target and acceptance limits. These figures show a linear desirability function ranging from zero (no desirable value for requirement) to one (most desirable response). These functions provide the building blocks for desirability calculations.

6.3   Desirability—single requirement

For the 3D scanning equipment example, four main functional requirement groups are identified (optical performance, portability and flexibility, reliability, and cost). For each requirement group, specific requirements including units of measure, direction of improvement, target values, and acceptance limits are defined. From this example, we first consider estimation of a desirability score for a single requirement.

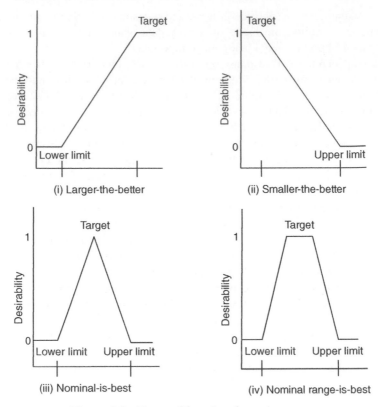

Figure 6.3   Types of functional requirements.

Calculations and optimization are distinguished by the direction of improvement type (one-sided versus two-sided limits), and the assumption regarding the shape of the desirability function (linear versus nonlinear). These conditions are explored in the following sections.

6.3.1   Desirability—one-sided limit

For smaller-the-better and larger-the-better desirability optimizations, we need a target value and a single acceptance limit. From here, we shall determine a desirability index score ranging from zero (not desirable) to one (maximum desirability). This range from zero to one measures how desirable it is for a response to take on a particular value within the acceptance region for the requirement.

To demonstrate desirability calculations, we shall use our scanning equipment example. One of the requirements for a scanner system is reliability. Suppose the organization measures this response using an accelerated test where the response metric is mean-time-until-failure in hours. The organization has an objective of

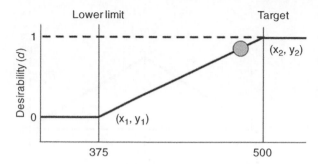

Figure 6.4   Larger-the-better desirability.

larger-the-better with a target of 500 hours and a lower acceptance limit of 375 hours. Figure 6.4 illustrates this larger-the-better case for an assumed linear desirability function.

To estimate a desirability score for this case, we may use Equation 6.1.

$$d_{\text{LTB}} = \begin{cases} 0 & \text{if actual} \leq \text{lower} \\ \left(\dfrac{\text{actual} - \text{lower}}{\text{target} - \text{lower}} \right) & \text{if lower} \leq \text{actual} \leq \text{target} \\ 1 & \text{if actual} \geq \text{target} \end{cases} \qquad (6.1)$$

If the actual value of the design solution is greater than or equal to the target in the larger-the-better case, we assign a desirability of one, indicating the desired requirement response is fully achieved. No additional desirability is awarded for values greater than the target value. As such, if the performance is much greater than the target, this may suggest an over-design condition and a potential opportunity to trade off with another underperforming requirement. In contrast, if the actual value is below the lower acceptance limit, we would assign a desirability of zero. In other words, not only is the actual value deviating from the desired target, but also it is outside our acceptance limits and thus may be viewed as a design solution having no value.

For this reliability requirement, suppose the company tests several products. From their tests, they estimate the actual mean-time-to-failure is 495 hours, slightly below their target but within their acceptance limits. Using the above larger-the-better equation, this would equate to a desirability score of 0.92 for reliability.

$$d_{\text{mean-time-to-failure}} = \left(\frac{490 - 375}{500 - 375} \right) = 0.92$$

Another requirement for the scanner system is optical head weight (measured in kilograms). Here, the direction of improvement is smaller-the-better with a target of 9 kg and an upper acceptance limit of 12 kg. Now, suppose in the initial

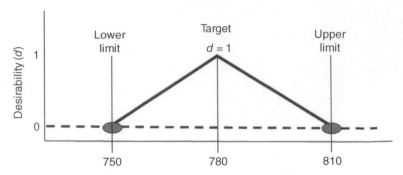

Figure 6.5   Nominal-is-best desirability.

design release, the manufacturer estimates the actual system weight of the product at 11 kg. We may compute a desirability score given a target value along with an estimate for the actual performance, assuming a linear objective function as defined by Equation 6.2.

$$d_{\text{STB}} = \begin{cases} 0 & \text{if actual} \geq \text{upper} \\ \left(\dfrac{\text{upper} - \text{actual}}{\text{upper} - \text{target}}\right) & \text{if target} \leq \text{actual} \leq \text{upper} \\ 1 & \text{if actual} \leq \text{target} \end{cases} \qquad (6.2)$$

For this scanner system, we would estimate a desirability score of 0.33 for optical head weight. This lower score could drive designers to consider alternative designs to improve performance of this metric.

$$d_{\text{system weight}} = \left(\frac{12 - 11}{12 - 9}\right) = 0.33$$

6.3.2   Desirability—two-sided limit

For some requirements, the objective is to hit a target value between two acceptance limits. Figure 6.5 illustrates an example with a two-sided limit for the requirement, *working distance.*[3]

To compute the desirability score for this condition, we may simply extend the smaller-the-better and larger-the-better cases by considering the relative position of the actual performance value relative to the target. Equation 6.3 provides an

---

[3] This value represents the desired distance between the optical head and the object being measured.

estimate of desirability for the nominal-is-best condition assuming a linear desirability function.

$$d_{NTB} = \begin{cases} 0 & \text{if actual} \geq \text{upper or actual} \leq \text{lower} \\ \left(\dfrac{\text{upper} - \text{actual}}{\text{upper} - \text{target}} \right) & \text{if target} \leq \text{actual} \leq \text{upper} \\ \left(\dfrac{\text{actual} - \text{lower}}{\text{target} - \text{lower}} \right) & \text{if lower} \leq \text{actual} \leq \text{target} \end{cases} \tag{6.3}$$

Suppose the working distance target requirement is equal to 780 mm with lower and upper acceptance limits of 750 and 810, respectively. If the initial design results in an actual value of 785 mm, this would be slightly above the target. Using the nominal-is-best equation, the desirability score for this requirement is 0.833.

$$d_{\text{working distance}} = \frac{810 - 785}{810 - 780} = 0.833$$

While not shown here, estimating desirability for nominal range is best may be easily derived from the above. The key difference is one has an additional range in the center which the actual value may fall between to obtain a desirability of one.

6.3.3   Desirability—nonlinear function

In some cases, one may wish to use a nonlinear desirability function. For instance, we may wish to put more, or less, emphasis on hitting the target value. To account for this, we may include a weighting factor to the desirability function. Here, the weight rating would change the shape of the desirability function. Figure 6.6 shows some commonly used weights for optimizing a desirability function.

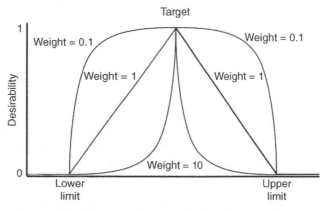

Figure 6.6   Weight values for desirability calculations.

Given these weights, we may express a more generic equation for desirability (Derringer & Such, 1980).[4] These equations are shown below for each of the cases.

Weighted desirability—larger-the-better

$$
d_{\text{LTB}} = \begin{cases} 0 & \text{if actual} \leq \text{lower} \\ \left(\dfrac{\text{actual} - \text{lower}}{\text{target} - \text{lower}} \right)^{\text{weight}} & \text{if lower} \leq \text{actual} \leq \text{target} \\ 1 & \text{if actual} \geq \text{target} \end{cases} \tag{6.4}
$$

Weighted desirability—smaller-the-better

$$
d_{\text{STB}} = \begin{cases} 0 & \text{if actual} \geq \text{upper} \\ \left(\dfrac{\text{upper} - \text{actual}}{\text{upper} - \text{target}} \right)^{\text{weight}} & \text{if target} \leq \text{actual} \leq \text{upper} \\ 1 & \text{if actual} \leq \text{target} \end{cases} \tag{6.5}
$$

Weighted desirability—nominal-is-best

$$
d_{\text{NTB}} = \begin{cases} 0 & \text{if actual} \geq \text{upper or actual} \leq \text{lower} \\ \left(\dfrac{\text{upper} - \text{actual}}{\text{upper} - \text{target}} \right)^{\text{weight}} & \text{if target} \leq \text{actual} \leq \text{upper} \\ \left(\dfrac{\text{actual} - \text{lower}}{\text{target} - \text{lower}} \right)^{\text{weight}} & \text{if lower} \leq \text{actual} \leq \text{target} \end{cases} \tag{6.6}
$$

For the above weighted desirability equations, a weight value of one reduces the equations to a linear function. A weight value less than one reduces the emphasis of hitting the target. For example, we could assign a weight between 0.1 and 1.0. A common sports analogy for such a condition is the "goal post-mentality." In sports with a goal, one is typically rewarded the same points whether the actual performance is exactly on target (center of the goal) or just inside the goal posts.

In contrast to weight values less than one, one may wish to increase the emphasis on hitting a target. Here, a weight value between 1 and 10 may be used. With a higher weight rating (e.g., weight = 10), a system designed with even a relatively small deviation from target would be viewed as a significant design concern.

In returning to our scanning example, suppose we have an acquisition speed requirement of 1 ± 0.25 s. This system may only need to be within its boundaries to obtain accurate, reliable measurements. As an aside, at this level of speed, a customer will certainly not notice a difference in terms of its cycle time performance. However, acquisition speed time can affect flexibility to different usage environments, for example, a factory floor versus a metrology room.

---

[4] In the original work on desirability by Harrington (1965), he proposed using an exponential function. However, the transformation function was less flexible in modeling different distribution shapes.

For this metric, suppose the initial design solution has an average acquisition speed at 1.05, which is slightly worse than the target. If we assign a low weight rating (0.1) for this metric, this deviation may be viewed as relatively insignificant. With a weight rating of 0.1, the desirability score would be equal to 0.91.

$$d_{\text{aquisition speed}} = \left[\frac{1.25 - 1.05}{1.25 - 0.75} \right]^{0.1} = 0.91$$

Of course, if the system was to deteriorate beyond the acceptance boundaries and leads to performance degradation, the desirability score would go to zero.

6.4   Desirability—multiple requirements

A key benefit of transforming single requirements into a zero-to-one desirability score is that requirements of different measurement units and scales may be converted to a uniformly scaled index. As such, we may transform a vector of different requirements or independent variables into a single total desirability index.

We now discuss several techniques to do this. First, we explore combining multiple responses for a *requirements group*. For instance, in the scanning system, we may have four different requirement groups (optical performance, portability and flexibility, reliability, and cost). Each of these groups may have a set of requirements affecting its performance. For example, optical performance may be represented by a combination of functional requirements such as field of view, working distance, and image acquisition. To determine the total desirability for a group of k requirements, where each requirement is assumed independent, and of equal importance, of the others, we may estimate total desirability by computing the geometric mean of the individual requirements using Equation 6.7. A rationale for using the geometric mean (versus say the arithmetic mean) is that if any individual requirement is deemed unacceptable ($d_i = 0$), then the entire design solution is deemed unacceptable.

$$D_{\text{system}} = (d_1 \times d_2 \times \cdots \times d_k)^{1/k} \tag{6.7}$$

The practical implication of having a total desirability index equal to zero, if any one individual requirement is zero, warrants further discussion. In establishing a design scorecard, one should exercise caution in including stretch goals or noncritical requirements. Here, one might have a solution whose actual performance is outside its limit yet does not result in a design with no value. In this situation, the issue lies with the establishment of appropriate acceptance limits that drive decision making.

The above estimation method assumes that each requirement is of equal importance to the overall desirability of the requirements group. In practice, organizations often place different levels of importance to various requirements. Derringer (1994) suggests a new form of the total desirability index using a weighted geometric mean to address this concern. This weighted total desirability index is shown in

Equation 6.8. Of note, if the weight values for each individual requirement are the same, this equation simplifies to the above total desirability index assuming equal importance.

$$D_{\text{system}} = \left[\prod_{j=1}^{k} \left(d_j^{i_j} \right) \right]^{\left(1/ \sum_{j=1}^{k} i_j \right)} \tag{6.8}$$

where i_j is the importance of the jth requirement.

To illustrate the weighted total desirability index, Table 6.2 summarizes the desirability scores and importance ratings for each requirement within the optical performance group.

The total desirability index for this requirement group, optical performance, may be estimated as

$$D_{\text{optical performance}} = \left(0.97^{4.7} \times 0.83^4 \times 0.95^{4.5} \right)^{\left(\frac{1}{4.7+4+4.5} \right)} = 0.92$$

Several different techniques may be used to establish importance ratings. For some organizations, the product develop team establishes them based on expert opinion and historical experience. Other organizations may obtain these based on customer data derived from surveys, interviews, or other methods of obtaining the voice of the customer.

In establishing importance ratings for different requirements, the relative differences among importance ratings often are more critical than the absolute ratings. For instance, a relatively low importance rating for a particular requirement could significantly dampen the effect of the individual requirement on the overall desirability score, even if its individual score is quite low. As such, one must be careful in considering the relative differences among importance ratings to avoid a misleading total desirability. As a general guideline, one should assume equal importance ratings among requirements unless they have solid evidence to suggest otherwise. Table 6.3 summarizes the desirability for each individual requirement and its group.

Total desirability for a set of requirements within a requirements group may be easily extended to a total desirability index for a complete design solution (multiple groups). Here, one simply replaces the desirability scores for individual requirements with group desirability scores.

$$D_{\text{system}} = \prod_{m=1}^{n} \left(d_m^{i_m} \right)^{1/ \sum_{m=1}^{n} i_m} \tag{6.9}$$

where m is the number of requirements groups, d_m is the desirability of the mth group, and i_m is the importance of the mth group.

Assigning an importance rating for an entire requirements group is different than assigning one for an individual requirement. Here, one is trying to assign a relative weighting factor to place greater emphasis on whatever is deemed as the

Table 6.2 Desirability scorecard[5] for optical performance group.

Item	Detail Requirements	Units	Importance	Weight	Desirability	Improvement Direction	Target	Lower Limit	Upper Limit	Actual
1.1	Field of view	cm^2	4.7	1.0	0.94	Larger-is-better	2600	900		2500
1.2	Working distance	mm	4.0	1.0	0.83	Target-is-best	780	750	810	785
1.3	Image acquisition	ms	4.5	0.1	0.95	Target-is-best	1	0.75	1.25	1.1
				Total desirability	**0.91**					

[5]Scorecard templates and desirability calculations were done using QETools© Excel® Add-In Software (qetools.com).

Table 6.3 Desirability scorecard for all requirement groups.

Item	Detail Requirements	Units	Importance	Weight	Desirability	Improvement Direction	Target	Lower Limit	Upper Limit	Actual
1 Optical performance										
1.1	Field of view	cm²	4.7	1.0	0.94	Larger-is-better	2600	900		2500
1.2	Working distance	mm	4.0	1.0	0.83	Target-is-best	780	750	810	785
1.3	Image acquisition	ms	4.5	0.1	0.95	Target-is-best	1	0.75	1.25	1.1
				Total desirability	**0.91**					
2 Portability and flexibility										
2.1	Optical head weight	kg	4.5	1	0.33	Smaller-is-better	9		12	11
2.2	Optical head size	cm³	4.0	1	0.60	Target-is-best	40 000	39 500	40 500	40 200
2.3	Cord length	meters	3.0	1	1.00	Target-is-best	4.5	4.0	5.0	4.5
2.4	Shipping crate size	cm³	3.5	1		Smaller-is-better				
				Total desirability	**0.54**					
3 Reliability										
3.1	Mean-time-to-failure	hours	3.7	1	0.92	Larger-is-better	500	375		490
				Total desirability	**0.92**					
4 Cost										
4.1	# Parts—bill of material	Count	3.5	1	0.93	Smaller-is-better	250		400	260
4.2	System cost	$	4	1	0.80	Smaller-is-better	200k		300k	220k
				Total desirability	**0.86**					

Table 6.4   Desirability scorecard – summary of multiple requirement groups.

Item	Requirements Group	Importance	Desirability Goal	Desirability Index
1	Optical performance	4.5	0.9	0.91
2	Portability and flexibility	4.5	0.9	0.54
3	Reliability	4.0	0.9	0.92
4	Cost	4.0	0.9	0.86
			Total desirability index	**0.78**

more *important requirements group* to satisfy. Table 6.4 summarizes the importance and desirability scores for each requirements group of the 3D scanning system. We also may include a benchmark goal in this scorecard for each requirements group to use in evaluating potential trade-offs.

Once we establish an importance rating for each group and its desirability score, we may compute a system total desirability index. The total desirability index for the 3D scanning system is 0.78.

$$D_{\text{system}} = (0.94^{4.5} \times 0.54^{4.5} \times 0.92^4 \times 0.86^4)^{\left(\frac{1}{4.5+4.5+4+4}\right)} = 0.78$$

The above information is often displayed using a radar chart as shown in Figure 6.7. With this chart, one may quickly see that the current design solution is lacking in

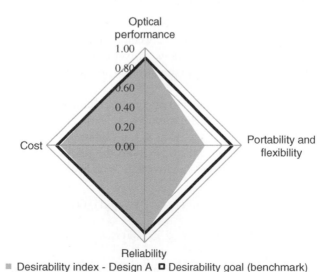

■ Desirability index - Design A ▢ Desirability goal (benchmark)

Figure 6.7   Radar chart – 3D scanning system.

Table 6.5   Desirability scorecard – comparing design alternatives.

Item	Requirements Group	Importance	Desirability Goal	Desirability Index Design A	Desirability Index Design B
1	Optical performance	4.5	0.9	0.91	0.90
2	Portability and flexibility	4.5	0.9	0.54	0.92
3	Reliability	4.0	0.9	0.92	0.90
4	Cost	4.0	0.9	0.86	0.82
	Total desirability index			**0.78**	**0.89**

portability and flexibility relative to the design goal. Unfortunately, improving the portability and flexibility score will result in some desirability loss for another group. For instance, one only may be able to address this deficiency through portability enhancements that might result in higher costs or reductions in optical performance. At this point, we often need to get creative to search for ways to improve the system without substantially hindering the performance of other requirements.

Another useful application of the total desirability index is to compare different design alternatives. Suppose the 3D scanning system is redesigned to address the portability concern. Here, we may use total desirability indices to compare and potentially select among various design options. An example of this is shown in Table 6.5.

6.4.1   Maxi-min total desirability index

One concern with total desirability indexes is their assumption of independence. There may be significant dependencies between requirement metrics related to the same system input factors. Unfortunately, early in product development, in cases where engineering functions are approximate or do not exist, these dependencies and relationships may not be known.

By assuming independence of the requirements (either using weighted or unweighted geometric mean), the total desirability index may be misleading (Khuri & Conlon, 1981). To address this issue, Kim & Lin (2000) propose using a maxi-min method. Here, rather than take a geometric average, the goal is to find a set of X factors which maximizes the minimum desirability score with respect to all responses.

$$D_{\text{Maxi-min Total}} = Maximize(\min[d_1\{\hat{y}_1(x)\}, d_2\{\hat{y}_2(x)\}, \dots \ d_r\{\hat{y}_r(x)\}])$$
$$x = \Omega$$
$$0 \leq d_j\{\hat{y}_j(x)\} \leq 1 \qquad j = 1, 2, \dots r \text{ responses}$$
$$(6.10)$$

One advantage of the above transfer method is that it may be used when two or more alternatives are being considered. Here, rather than create separate indexes for each design option, one value may be obtained. Still, the maxi-min approach does have disadvantages. Most notably, by focusing on the requirement with the lowest desirability, all other requirements are effectively ignored. As such, even in the presence of moderate dependencies among requirements, we recommend the use of the total desirability index for most situations.

6.5   Desirability—accounting for variation

The above estimates of desirability have been based on nominal estimates, typically the mean, to summarize performance. In rare cases, this assumption may be reasonable if the metric may be set to a specific value with minimal variation.

Ignoring the variation may result in a poor characterization of a design's desirability. For instance, if the nominal estimate of the mean is based on a small sample of product (which is fairly common with prototypes and early production builds during new product development), it may not reflect the true value of the design. Next, we discuss ways to address this concern.

When engineering models are not available or require an excessive number of assumptions, one may consider power and sample size planning (including Type I and II errors) to develop a sufficient sample size in which to compute mean estimates with acceptable error levels (Montgomery & Runger, 2011). While this approach reduces the variability in the estimated mean, the inherent variability of the output is ignored. Another difficulty of this approach is the difficulty of obtaining the required number of samples.

A core objective of Design for Six Sigma is to make response outputs as insensitive as possible to variation in process input variables (Yang & El-Haik, 2003). To assess robustness, several metrics have been proposed. One of these is the signal-to-noise ratio (Taguchi et al., 2005). This metric is not recommended as it does not allow the desired target to be reached while simultaneously minimizing variation. A large variation is tolerated as long as the signal is high. Of course this can be overcome by setting the signal to the desired target level, but it is easier to simply minimize variation.

Since desirability scores may be viewed as simply a transformation of performance relative to acceptance limits bounded between zero and one, one may replace the actual average with any measure of performance (signal-to-noise ratio, variance, 95th percentile, etc.). When doing this, the target and acceptance limits likely would need to be transformed to an appropriate scale to match the metric of interest. For instance, upper and lower bounds on the variance would certainly be different than upper and lower bound on the mean. In this section, we present additional techniques to account for variation in assessing desirability.

6.5.1   Determining desirability—using expected yields

To account for processing variation, one approach is to use the expected yield in place of the desirability score. Here, we may view the production of individual units within their assigned technical specifications as desirable (good) and those outside their specification limits as undesirable (defective). In doing so, we may compute the expected yield of a process. Since this expected yield will range from zero (none in specification) to one (all parts within specification), we may use this in place of the desirability score for a particular requirement. The value in using this approach is that we are assigning greater desirability for outputs that both achieve a desired mean value and have sufficiently low variation to meet product specification limits.

To illustrate this approach, suppose we measure the image acquisition speed for a sample of 50 trials. The results are shown in the process capability graphical summary shown in Figure 6.8.[6] The estimated process mean is 1.05 with a sample standard deviation of 0.091. Assuming a normal distribution and specification limits of 0.75 and 1.25, we may estimate the probability that this process will produce an acceptable product.

For this example, the expected yield is estimated as 98.5%, which becomes the desirability score for this requirement, *image acquisition speed* (see Ribardo & Allen, 2003 for a further discussion on the use of yields to measure desirability). If the assumed distribution is non-normal, we may adjust our expected yield estimates using the techniques described in Chapter 5. Bothe (1997) also discusses desirability with non-normal metrics.

6.5.2   Determining desirability—using non-mean responses

Rather than using expected yields to account for processing variation, we may directly compare variation or percentiles to a desired target value with appropriate limits. For instance, suppose we wish to design a product to achieve a specific level of variation. In the case of the image acquisition, suppose there is a target for standard deviation of 0.05 with an upper limit of 0.10 (standard deviation would typically be optimized using lower-the-better). If the actual process standard deviation is estimated at 0.091, the desirability score assuming a linear objective function would be 0.18.

$$d_{\text{image speed std dev}} = \frac{0.10 - 0.091}{0.10 - 0.05} = 0.18$$

From this perspective, the standard deviation may be viewed as relatively undesirable. Again, we should point out the potential dangers of using point estimates for desirability estimates. Standard deviation should be estimated using engineering models along with the error propagation methods described in Chapter 5, whenever possible. If engineering models do not exist, or there is an abundance of assumptions required when using engineering models, then samples may be required to estimate variation. In this case, with a sample of only 50 units, the 95% confidence intervals for our standard deviation estimate would range from 0.077 to 0.114. As such, we may wish

---

[6] This process capability graphical summary was obtained using QETools Statistical Software, though numerous other statistical quality analysis software packages provide similar output.

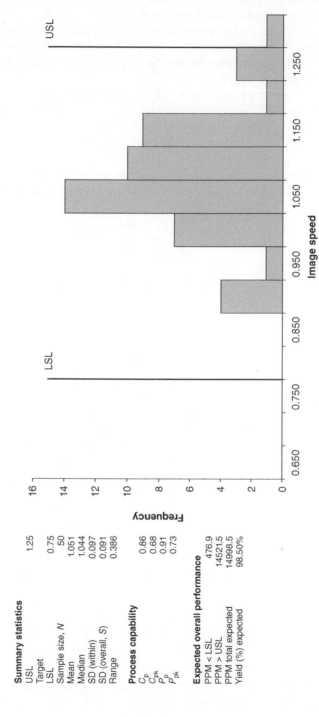

Figure 6.8   Process capability graphical summary – image acquisition speed.

to use larger sample sizes or perhaps theoretical values for design simulations as more appropriate predictions of actual performance.

A final way to account for variability of expected output is to use a "worst-case" estimate. Here, we may first determine the 90th or 95th percentile of our response distribution. We may then compare this value to our original acceptance limits or to different limits for this statistic.

Given that we rarely have acceptance limits for say a 95th percentile measurement, we likely would use the original acceptance limits. In this case, we would expect low desirability scores. Still, we may use these estimates to provide a potential performance range from a typical unit (average) to an extreme or worst-case condition (95th percentile).

For this example, if the 95th percentile image acquisition speed is estimated at 1.23, the desirability score assuming a linear objective function would be 0.08.

$$d_{\text{image speed 95th}} = \frac{1.25 - 1.23}{1.25 - 1.0} = 0.08$$

Given a desirability of 0.08 based on a mean estimate of 1.25, we can see that the desirability could vary significantly depending on the effect of variation.

6.6   Summary

Developing balanced designs are critical to meeting the needs of a diverse customer base with multiple requirements. The use of design scorecards provides an effective tool to convey these requirements and their expected performance. These scorecards become even more powerful when converted into a measure of desirability. This allows designers to consider trade-offs and identify the best overall solution based on the information available.

The literature contains many methods for including variability in performance metrics, but nearly all of these metrics are based on physical measures of multiple products. This is not only time-consuming and expensive, but also it is difficult to get an accurate assessment of performance because of small sample sizes and the inability to include all sources of variation in the parts tested. The method of estimating the output variation by propagating input variation introduced in Chapter 5 is the most efficient and accurate method available. This method will be expanded in the remaining chapters.

Exercises

1   The article "Methodology for Examining the Simultaneous Impact of Require-ments, Vehicle Characteristics and Technologies on Military Aircraft Design"

(Mavris & DeLaurentis, 2000) takes a deep dive into the design of a military aircraft and reports the following nine functional requirements which are to be optimized for a good aircraft design. Based on these requirements, the engineers at an aerospace company have developed a design with the values shown in the "actual value" column of the table.

Requirement	Direction of Improvement	Acceptance Criteria	Importance	Actual Value
Mission radius (nm)	Nominal-is-best	Nominal = 365.5 Min = 296 Max = 435	5.0	405
Ultimate load factor (no unit)	Nominal-is-best	Nominal = 7.2 Min = 6.5 Max = 7.9	4.5	6.7
Combat Mach number (no unit)	Nominal-is-best	Nominal = 1.0 Min = 0.9 Max = 1.1	4.0	0.99
Mission payload (lb)	Smaller-the-better	Max = 1000 Target = 500	3.0	700
Thrust per engine (lb)	Nominal-is-best	Nominal = 17 750 Min = 14 500 Max = 21 000	5.0	17 750
Ref. Wing Area (ft^2)	Nominal-is best	Nominal = 450 Min = 380 Max = 520	5.0	430
Stealth penalty (lb)	Smaller-the-better	Max = 1000 Target = 700	4.5	500
Auxiliary fuel tanks (no. of tanks)	Nominal-is-best	Nominal = 1 Min = 0 Max = 2	3.0	1
Specific fuel consumption; k-factor	Nominal-is-best	Nominal = 0.95 Min =0.9 Max = 1	3.0	0.93

(a) Determine the desirability for the requirement, Ref. Wing Area?

(b) Determine the overall desirability of the aircraft which the engineers have designed?

2 The given data on the requirements of JS-SS400 type of steel are taken from the paper "Simultaneous Optimization of Mechanical Properties of Steel" (Kim & Lin, 1999). An American steel manufacturer produces this steel for various automobile suppliers. Among the three responses, Y1 is considered the most important, followed by Y2 and then Y3. The manufacturer's market research team also agreed on employing a convex, linear, and concave desirability function for YI, Y2, and Y3 with weights of 0.3, 1.0, and 3.0, respectively.

Bounds and Target	Requirements		
	Hardness (Y1)	Cohesiveness (Y2)	Springiness (Y3)
Minimum	43.10	26.30	20.00
Maximum	52.00	–	–
Target	47.55	48.08	43.67
Actual value	49.10	44.38	30.16
Direction of improvement	Nominal-is-best	Larger-the-better	Larger-the-better
Importance	5	4	3
Weight	0.3	1.0	3.0

(a) The manufacturer currently produces steel with the specifications shown in the "actual value" row of the table. Calculate the desirability of the steel being produced by the manufacturer?

(b) The manufacturer wants to improve its market share by increasing its product desirability. Due to financial limitations, the company can allot funds to improve one requirement only. Among the three requirements, which one should the company concentrate on improving, to have the highest increase in the product desirability? What is the new product desirability when the target is achieved for this requirement?

3 Consider a plastic material whose quality is defined by five principle properties as follows.

Requirement	Properties	Max	Min	Target	Direction of Improvement	Importance	Actual Value	Desirability Goal
Strength of the material	Tensile strength (psi)	–	7000	10 000	Larger-the-better	5	9800	0.9
	Tensile modulus ($\times 10^5$)	5.5	4.5	5	Target-is-best	3	4.67	
Temperature resistance	Deflection temp. (°F)	–	190	270	Larger-the-better	3	205	0.8
	Dielectric constant	2.60	–	1.50	Smaller-the-better	3	1.70	
Manufactur-ability	Flow index	1.40	1.20	1.30	Target-is-best	4	1.30	0.9

(a) Calculate the desirability for the requirement "Strength of the material"?

(b) Draw a "Desirability radar chart." Compare the current material with the benchmarked material and list the requirements that need to be improved.

4 Consider an electronic dispenser used for dispensing polyamide coating material in various electronic manufacturing processes. One of the requirements of the dispenser is that it should dispense 1000 mg of coating each time it is operated (requirement is amount dispensed). The amount dispensed was measured for a sample of 90 trials. The results are shown in Figure 6.9.

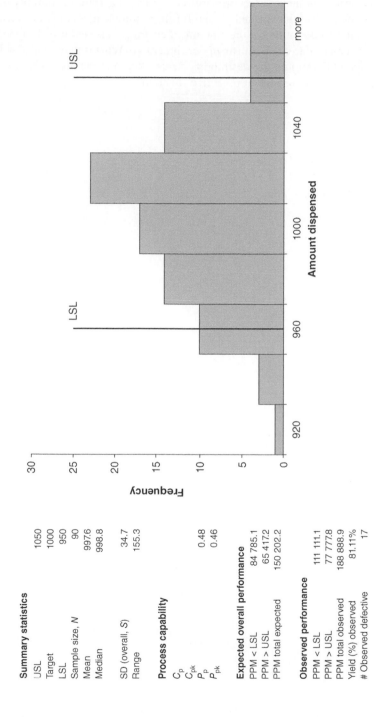

Figure 6.9 Electronic dispenser process capability.

(a) Given the following process capability analysis in Figure 6.9, assuming a normal distribution, and a defect probability of 0.15, what is the desirability score using the method of expected yields, for the requirement amount dispensed?

(b) The standard deviation for the amount of coating dispensed was designed to have a target value of 25 with an upper limit of 40. What is the actual standard deviation? Determine the desirability score using the method of non-mean responses assuming a linear objective function?

7

Optimization and sensitivity

Models for the nominal performance and variation can be used to optimize designs for robustness. When the number of inputs exceeds the number of outputs, multiple solutions exist, and the design engineer has the ability to create a design that minimizes the sensitivity of the outputs to variability of the inputs.

For example, if the design problem is to determine the angle and velocity required to make a projectile travels 2000 feet in a vacuum, then any point on the line shown in Figure 7.1 provides a solution. The goal is to choose the solution that yields minimum variation in the distance the projectile travels. The models introduced in Chapter 5 can be used with optimization routines to achieve the target output while minimizing the variance of the output.

7.1 Optimization procedure

The following procedure is recommended for optimizing a design for robustness.

1. Model nominal output corrected for statistical bias using the first two terms of the Taylor series approximation.

2. Model output variation using the first term of the Taylor series approximation.

3. If there are fixed output specifications, use an optimization algorithm to minimize the output standard deviation while achieving the desired nominal output.

4. If there are fixed output specifications, select the appropriate statistical distribution to model the output data, and use an optimization algorithm to maximize the process capability index (C_{pk}).

5. Verify the results with Monte Carlo simulation.

Probabilistic Design for Optimization and Robustness for Engineers, First Edition.
Bryan Dodson, Patrick C. Hammett and René Klerx.
© 2014 John Wiley & Sons, Ltd. Published 2014 by John Wiley & Sons, Ltd.
Companion website: http://www.wiley.com/go/robustness_for_engineers

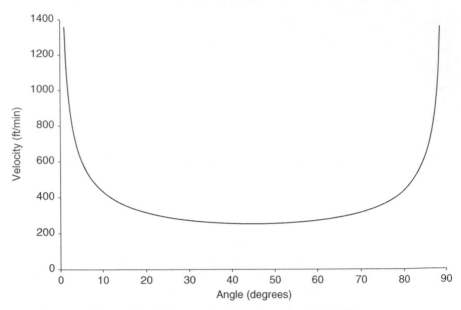

Figure 7.1   Solutions for firing a projectile 2000 feet.

6. If the results for output variance from the Taylor series approximation vary excessively from the simulation results:

   (a) model variance output variation using the second-order Taylor series approximation.

   (b) verify results with simulation.

It is recommended to begin the robustness process with the first-order approximation for variance because the first-order estimate provides reasonable accuracy much of the time, and the number of required terms for the second-order approximation increases dramatically with the number of system inputs.

Example 7.1 Polynomial robustness[1]

Given Equation 7.1, what nominal values of T and R yield a value of 7000 for Y while minimizing the variation of Y given the standard deviation of T is 10, and the standard deviation of R is 5?

$$Y = 25 + 200R - 2R^2 + 14T \tag{7.1}$$

---

[1] This example is contained on the accompanying website in both Excel® and Sage®. The spreadsheet file name is SecondOrderPoly.xls and the Sage® file names are SecondOrderPolynomialNotRobust.txt, and SecondOrderPolynomialRobust.txt.

Solution

Many engineers would choose a nominal value of zero for R, which simplifies the equation, allowing an easy solution for T.

$$T = \frac{7000 - 25}{14} = 498.21$$

By using the robustness procedure, the standard deviation of Y can be estimated as shown below.

$$\frac{\partial Y}{\partial R} = 200 - 4R$$

$$\frac{\partial^2 Y}{\partial R^2} = -4$$

$$\frac{\partial Y}{\partial T} = 14$$

$$\frac{\partial^2 Y}{\partial T^2} = 0$$

From Equation 5.6, the estimated bias is

$$\text{Bias} = \frac{1}{2}[(-4)(5^2) + (0)(10^2)] = -50$$

$$\sigma_Y = \sqrt{(200 - 4R)^2\,(5^2) + (14)^2\,(10^2)}$$

$$Y = 25 + 200(0) - 2(0)^2 + 14(498.21) - 50 = 6950$$

$$\sigma_Y = \sqrt{(200 - 4(0))^2\,(5^2) + (14)^2(10^2)} = 1010$$

The calculations above are summarized in Table 7.1.

The results of this method should always be verified with a simulation as shown in Figure 7.2.

The simulation verifies correct estimates of the nominal value of Y (6950) and the standard deviation of Y (1010). In this case, the traditional engineering approach

Table 7.1   Initial solution for estimated Y expected value and standard deviation.

Parameter	Nominal	Input Standard Deviation	First Derivative	Second Derivative	Output Variance	Output Standard Deviation	Bias
R	0	5	200	-4	1 000 000	1000	-50
T	498.2	10	14	0	19 600	140	0
Y (theoretical)	7000						
Y (actual)	6950						
Y (standard deviation)	1010						

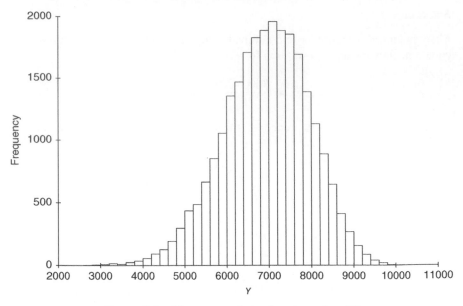

Figure 7.2 Histogram of simulated results of Y.

of using the simplest solution not only fails to provide the most robust solution, but also the target is missed (6950 vs. a target of 7000). To find the most robust solution, an optimization routine can be applied to minimize the standard deviation of Y while forcing the nominal value of Y, after correction for statistical bias, to be equal to 7000. This can be done using Solver feature in Excel® as shown in Figure 7.3.

	A	B	C	D	E	F	G	H
1	Parameter	Nominal	Input standard deviation	Y'	Y''	Output variance	Output standard deviation	Bias
2	R	0.0	5	200.0	−4	1000000.0	1000.0	−50
3	T	498.2	10	14	0	19600	140.0	0
4	Y (theoretical)	7000						
5	Y (actual)	6950						
6	Y - STD	1010						

Solver Parameters ⊠

Set Target Cell: $B$6   Solve

Equal To: ○ Max ⦿ Min ○ Value of: 0   Close

By Changing Cells:

$B$2:$B$3   Guess

Subject to the Constraints: Options

$B$5 = 7000   Add

Change   Reset All

Delete   Help

Figure 7.3 Solver solution for robustness of Y.

Table 7.2   Robust solution for Y.

Parameter	Nominal	Input Standard Deviation	First Derivative	Second Derivative	Output Variance	Output Standard Deviation	Bias
R	50.0	5	0.0	−4	0.0	0.0	−50
T	144.6	10	14	0	19 600	140.0	0
Y (theoretical)	7050						
Y (actual)	7000						
Y (STD)	140						

The robust solution for Y is shown in Table 7.2.

This solution with $R = 50$ and $T = 144.6$ estimates the standard deviation of Y to be 140. Monte Carlo simulation, shown in Figure 7.4, verifies the nominal Y is 7000, but estimates the standard deviation of Y to be 157. Referring to Table 7.2, the contribution of the parameter R to the standard deviation of Y is zero. According to the first-order approximation, the entire predicted standard deviation of Y is caused by variation in the parameter T. This is a common occurrence, as minimizing the standard deviation using the first-order approximation often finds points where the derivative is equal to zero. When R equals 50, the derivative of Y with respect to R is zero, and the estimated contribution to the standard deviation of Y is zero, regardless of how much parameter R varies. When using the first-order approximation for standard deviation, the correct optimum point is often found, but the standard deviation is

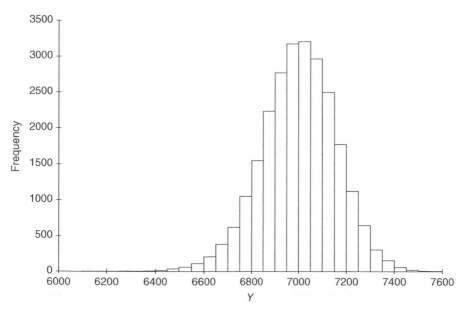

Figure 7.4   Histogram from Monte Carlo simulation of robust solution for Y.

usually underestimated. If the standard deviation estimate is relatively close to the result from simulation, there is no need to use the second-order term.

Using robust design for this problem reduced the standard deviation from 1010 to 157, a reduction of 84%, without reducing the variation of the inputs. Traditionally, if the variation of the outputs needs reduction, this is achieved by reducing the variation of all of the inputs. Robust design is a preferred method as reducing input variation often increases cost.

7.2   Statistical outliers

A statistical outlier is caused by a deviation from normal operation, such as a loss of power, drop in air pressure, lack of grease, etc. This special cause is beyond the output variation caused by normal variation of the inputs. In Figure 7.4, it can be seen that without special cause variation, Y varied from approximately 6000 to 7600 in 25 000 trials. This does not mean that no special cause variation is present when Y falls in this range, and it also does not mean that there is special cause variation if Y falls outside this range. If 10 million trials are simulated, Y would be expected to have a larger range as there would be more opportunities for R and T to align at the extremes of their statistical distributions.

When a value falls outside a significant distance from the expected value, for example, if $Y = 5900$, it is desirable to know if this is a result of common cause variation or special cause variation. Outliers can only be confirmed if there is a physical cause identified. A common statistical definition of an outlier is any value that is beyond 150% of the width of the inter-quantile range, from either end of the inter-quantile range. This is an unreliable method for identifying outliers. Figure 7.5 shows an outlier analysis from a commonly used commercial statistics package.

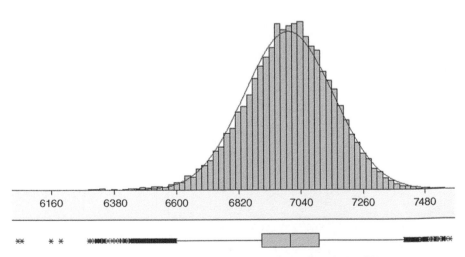

Figure 7.5   Statistical outlier analysis of optimized Y.

The asterisks represent statistical outliers. Note the large number of outliers. When verifying a design, human nature is to dismiss a poor value as an outlier and test another sample. This is dangerous, and it is not recommended to label a result as an outlier unless an engineering abnormality, such as a test aberration, measurement error, or other known cause for the value, is identified.

7.3   Process capability

If a process is normally distributed, the process capability index is

$$C_{pk} = \min \left[\frac{\text{upper spec limit} - \mu}{3\sigma}, \frac{\mu - \text{lower spec limit}}{3\sigma} \right] \qquad (7.2)$$

where μ is the distribution mean, usually estimated by the process average, and σ is the distribution standard deviation, usually estimated by the sample standard deviation. C_{pk} can be equated to the percentage of parts not meeting specifications. The percentage of parts not meeting the specification nearest to the distribution mean is

$$p = \Phi(-3C_{pk}) \qquad (7.3)$$

where $\Phi(x)$ is the standard normal cumulative distribution function.

The percentage of parts not meeting the nearest specification as a function of C_{pk} is shown in Figure 7.6.

During production, it is desirable to have a C_{pk} greater than 1.33. With a C_{pk} greater than 1.33, there is sufficient cushion between the tail of the distribution and the customer requirements so that no special production controls, such as 100% inspection, are required to prevent out-of-specification parts from being shipped to the customer. In the time period between the release of a design and the beginning of production, many things may change, and unknown and unexpected changes usually do not improve the design or the process. Because of this, when a final design is released it is desirable to have a C_{pk} greater than 1.67. Figure 7.7 displays a C_{pk} of 1.0, a C_{pk} 1.33, and a C_{pk} of 1.67.

The standard method of computing C_{pk} assumes a normal distribution as shown in Figure 7.7. If there is deviation from the normal distribution, the process may produce parts that do not meet specification even when the C_{pk} computed using Equation 7.2 is greater than 1.33.

In this case, the C_{pk} can be computed by determining the percentage of parts not meeting the weak specification (the specification nearest to the average) and computing the C_{pk} using the expression

$$C_{pk} = \frac{-\Phi^{-1}(p)}{3} \qquad (7.4)$$

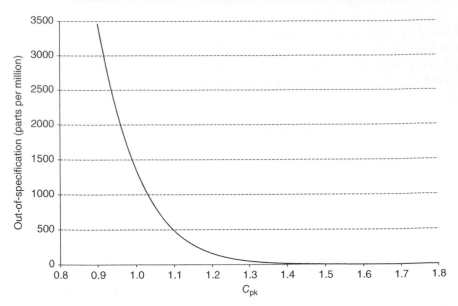

Figure 7.6   Percentage of parts not meeting nearest specification as a function of C_{pk}.

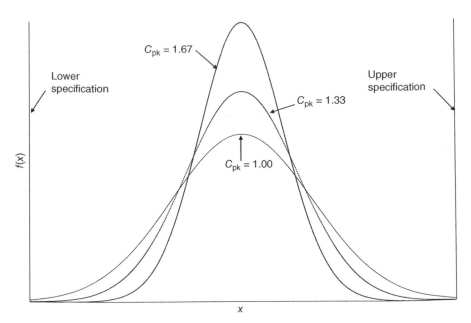

Figure 7.7   Process capability examples.

where $\Phi^{-1}(x)$ is the inverse of the standard normal distribution and p is the percentage of parts not meeting specification on the weak side.

The percentage of parts not meeting the weak specification can be determined by using the appropriate distribution or by counting the out-of-specification conditions from a Monte Carlo simulation.

Suppose the polynomial example defined by Equation 7.1 has a customer requirement of $6500 < Y < 7500$. Using the estimates for the mean and standard deviation of Y given in Table 7.2, and assuming a normal distribution for Y, the C_{pk} for Y is estimated as

$$C_{pk} = \min\left[\left(\frac{7500 - 7000}{3(140)}\right), \left(\frac{7000 - 6500}{3(140)}\right)\right] = 1.19$$

If the C_{pk} is included as a constraint in the optimization process, there would be a convergence failure because it is not possible to obtain a C_{pk} of 1.67 or even 1.33 given the inputs. Of course using the standard deviation obtained from the first-order Taylor series approximation when computing C_{pk} provides an optimistic estimate as the true standard deviation is underestimated. In this case, the C_{pk} is further overestimated because the output is not normally distributed. Using the second-order Taylor series approximation for standard deviation to compute C_{pk} provides a reasonable estimate for the standard deviation of the output, but deviation from normal will lead to error in the estimated C_{pk}. To ensure correct estimates, Monte Carlo simulation should be used for confirming the C_{pk} calculated from an estimated mean and standard deviation.

When using simulation to estimate C_{pk}, enough trials must be conducted to ensure a steady-state solution. As a rule of thumb, a minimum of 5000 trials is recommended. When computing out-of-specification conditions, several million trials may be required to achieve steady state. For example, if an out-of-specification condition occurs five times per million, 10 to 20 million trials are required to obtain an accurate estimate.

There are two options for using simulated output values to compute the C_{pk}:

1. fit a distribution to the output, use the fitted distribution to estimate the probability of not conforming to specifications, and compute the C_{pk} as described in Equation 7.4 and

2. count the simulated points not conforming to specifications and use this probability to compute C_{pk} as described in Equation 7.4.

In this example, it is not possible to fit a distribution to the Y values. The distribution is skewed to the left which eliminates the lognormal distribution from consideration. A Weibull probability plot for Y is shown in Figure 7.8. The concave curve of the plotted points indicates a poor fit.

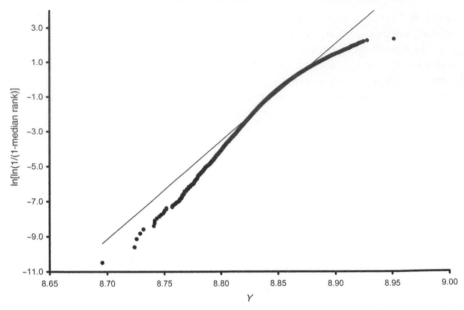

Figure 7.8   Weibull probability plot for Y.

The Monte Carlo simulation created in Excel® above was augmented with the following statements in two columns.

$$= IF(Y < 6500, 1, 0)$$
$$= IF(Y > 7500, 1, 0)$$

The first statement places a "1" in the cell if the simulated trial for Y is less than 6500 and the second statement places a "1" in the cell if the simulated trial for Y is greater than 7500. The average of each of these columns estimates the probability of falling below the lower specification and the probability of exceeding the upper specification. From a simulation consisting of 25 000 trials, the probability of falling below the lower specification is estimated at 0.00366 and the probability of exceeding the upper specification is estimated at 0.00032. Since the probability of falling below the lower specification is larger than the probability of exceeding the upper specification, the C_{pk} is determined by the probability of falling below the lower specification.

$$C_{pk} = \frac{-\Phi^{-1}(0.00366)}{3} = 0.89$$

It is important to correctly estimate the C_{pk} to ensure a product meets requirements before investments are made in tooling. At this point, the design is on paper, and changes can be made at relatively low costs to improve the design. The C_{pk} found using the standard deviation from the first-order estimate and assuming a normal distribution for Y is 1.19 (mean = 7000 and standard deviation = 140). The C_{pk} estimate using the standard deviation from the simulation but still assuming a normal distribution for Y is 1.06 (mean = 7000 and standard deviation = 157), while the best estimate for C_{pk} eliminates the normality assumption, reducing the estimated C_{pk} to 0.89 (C_{pk} computed from the probability of exceeding the upper specification).[2]

Maximizing C_{pk} provides better results than minimizing the standard deviation. Maximizing the C_{pk} allows the nominal output (after bias correction) to deviate from the center of the specifications. This allows the nominal of a right skewed distribution to deviate from the center of the specifications resulting in a reduced out-of-tolerance rate. Optimizing C_{pk} is especially useful for one-sided specifications as it is not clear where the nominal should be located if the goal is to minimize output standard deviation.

7.4   Sensitivity and cost reduction

After a design has been optimized for robustness, further improvement will be required if the process capability is not sufficient. The minimum recommended C_{pk} is 1.33, while a C_{pk} of 1.67 is preferred. Improvements can be made by reducing the standard deviation of the input parameters or by removing constraints. In some cases, the process capability will exceed requirements, and there will be opportunities to reduce costs by increasing the standard deviation of input parameters or changing nominal values to lower cost options.

Input standard deviation can be changed in many ways. For example, the standard deviation of electrical resistors can be reduced by using 1% resistors instead of 5% resistors. Surface finish standard deviation can be improved by honing rather than reaming. The number of parts produced from a tool between maintenance actions such as sharpening will change the standard deviation of grinding or cutting operations.

The traditional approach to reducing output variation is to reduce the variation of as many inputs as possible. In the case of the distance a projectile travels in a vacuum, the distance variance could be reduced by reducing the variance of the angle and the variance of the velocity. This is a costly and time-consuming approach. By using the sensitivities from the robustness model, the variance of the high sensitive parameters can be reduced to improve performance, while the variance of the insensitive parameters can be increased to reduce costs. There are three possible scenarios after a design has been optimized. The process capability can either be unacceptable, marginally acceptable, or acceptable with a safety margin. In all cases, actions should be taken to reduce costs. These scenarios are summarized in Table 7.3.

The concepts described above will be demonstrated with an example problem. This problem will build on the techniques introduced previously.

---

[2] This example is contained on the accompanying website. The spreadsheet file name is Second OrderPolyCpk.xls. The Sage® file name is SecondOrderPolynomialCpk.txt.

Table 7.3   Process capability scenarios.

Scenario	Action
C_{pk} unacceptably low	Reduce variance of highly sensitive parameters and increase the variance of insensitive parameters
C_{pk} marginally acceptable	Increase the variance of insensitive parameters
C_{pk} acceptable with a safety margin	Increase the variance of insensitive and sensitive parameters

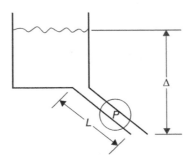

Figure 7.9   Water pumped from a reservoir.

7.4.1   Reservoir flow example

Figure 7.9 shows a reservoir with water being pumped through a pipe and discharged at a lower level. The engineer is responsible for achieving a pump horsepower[3] between 5 and 45 by determining the following design parameters:

- flow rate (v) in gallons per minute,

- inside pipe diameter (d) in inches,

- friction coefficient of pipe interior (f),

- height difference between reservoir level and discharge level (Δ) in feet, and

- pipe length (L), which is fixed at 2000 feet.

The design constraints and standard deviations of the inputs are given in Table 7.4. All inputs are normally distributed.

---

[3] In reality horsepower would be an input and the engineer would be designing a system to achieve a desired flow rate. The problem has been modified to determine horsepower because the flow rate equation is complicated with derivatives having an excessive number of terms.

Table 7.4   Reservoir flow example constraints.

Parameter	Lower Constraint	Upper Constraint	Standard Deviation
Flow rate (v)	10	100	2
Inside pipe diameter (d)	0.5	10	0.1
Friction coefficient of pipe interior (f)	0.01	0.05	0.002
Height difference between reservoir level and discharge level (Δ)	5	1000	3
Pipe length (L)	2000	2000	1

	A	B	C	D	E	F	G	H	I	J
1	Design parameter	Design nominal	Design constraint (lower)	Design constraint (upper)	Parameter standard deviation	h'	h''	Variance contribution	Standard deviation contribution	Bias
2	v	26.2	10	100	2	15.8	1.7	1003.0	31.7	3.44
3	d	4.7	0.5	10	0.1	-32.2	-15.8	10.4	3.2	-0.08
4	f	0.010	0.01	0.05	0.002	19271.2	0.0	1485.5	38.5	0.00
5	Δ	498.7	5	1000	3	-0.4	0.0	1.1	1.1	0.00
6	L	2000	2000	2000	1	0.1	0.0	0.0	0.1	0.00
7	h (theoretical) =	21.6								
8	h (actual) =	25.0								
9	h (STD) =	50.0								

Figure 7.10   Initial solution for the reservoir flow example.[4]

The equation below gives the pump horsepower as a function of the input parameters.

$$h = 0.00000961\, v^3 d^2 + 0.000115\, v^3 dfL - 0.0006188\, vd^2 \Delta \qquad (7.5)$$

7.4.2   Reservoir flow initial solution

Figure 7.10 provides an initial solution for the reservoir flow design in spreadsheet format.

The average horsepower is centered between the specification limits and all design constraints for the design parameters are satisfied. The standard deviation of the pump horsepower is 50 which, assuming a normal distribution, yields a process capability index of

$$C_{pk} = \min\left[\frac{45-25}{3\,(50)}, \frac{25-5}{3\,(50)}\right] = 0.13$$

---

[4]This example is included on the accompanying website. The file name is ReservoirHorsePower.xls. There is also a solution programmed in Sage®. The Sage® file name is ReservoirFlowInitialSolution.txt.

Instead of assuming a normal distribution for horsepower, a simulation should be performed to verify the horsepower distribution, however, a C_{pk} of 0.13 is so poor the distribution used for computing C_{pk} is not important at this point in the solution. Even though the design is not acceptable, the modeling itself provides value because it is better to discover a poor design early. Often poor capability is not discovered until production begins. This results in expensive re-design, scrap, re-work, and unhappy customers.

7.4.3   Reservoir flow initial solution verification

After each design iteration, it is advisable to create a Monte Carlo simulation for horsepower. The simulation provides three purposes:

- a verification against computational errors,

- an accuracy check of the first-order variance approximation, and

- a method for determining the statistical distribution of the horsepower.

The following expressions can be added to the spreadsheet above to obtain a Monte Carlo simulation for horsepower (Table 7.5).

Cell H11 will be "1" if the horsepower is less than the minimum specification, and "0" if horsepower is greater than the maximum specification. The average of column H yields the percentage of an out-of-specification condition for low horsepower. Column I repeats this analysis for the upper specification, and the average of column J is the estimated percentage not meeting either the lower or upper specification.

To reach a steady-state solution for the simulation, row 11 in the spreadsheet should be copied down at least 5000 times. The number of trials required to reach a steady-state simulation is dependent upon the system being simulated and the magnitude of the standard deviations of the system inputs. A histogram of horsepower from the Monte Carlo simulation is shown in Figure 7.11.

The mean and standard deviation output from the simulation which matches with those found from the statistical model is displayed in Figure 7.10, validating the

Table 7.5   Monte Carlo simulation expressions for horsepower.

Cell	Expression
B11	=E$2*((COS(2*PI()*RAND()))*(-2*LN(RAND())))^0.5)+B$2
C11	=E$3*((COS(2*PI()*RAND()))*(-2*LN(RAND())))^0.5)+B$3
D11	=E$4*((COS(2*PI()*RAND()))*(-2*LN(RAND())))^0.5)+B$4
E11	=E$5*((COS(2*PI()*RAND()))*(-2*LN(RAND())))^0.5)+B$5
F11	=E$6*((COS(2*PI()*RAND()))*(-2*LN(RAND())))^0.5)+B$6
G11	=$M$2*B11^3*C11^2+$M$3*B11^3*C11*D11*F11-$M$4*B11*C11^2*E11
H11	=IF(G11<5,1,0)
I11	=IF(G11>45,1,0)
J11	=SUM(H11:I11)

The calls $M$2, $M$3 and $M$4 contain the constants from equation 7.5.

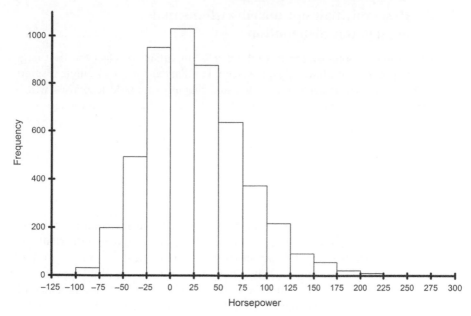

Figure 7.11   Histogram of initial reservoir flow solution.[5]

computations and accuracy of the first-order approximation for horsepower standard deviation. The histogram created from Monte Carlo simulation shows the horsepower is not normally distributed. This is to be expected as the horsepower is a result of the multiplication of variables. The histogram is right skewed, which is consistent with the behavior of a lognormal distribution. Further evidence is needed to confirm a lognormal distribution, usually a lognormal probability plot. Validation of the statistical distribution is not required until a final design is reached, however, when optimizing it is beneficial to use the correct statistical distribution. If the distribution is symmetrical, the optimum solution will place the mean in the center of the specification window. If the distribution is skewed right, as in this case, the optimum solution will place the mean closer to the lower specification. Keep in mind that the distribution may change as the optimization process reduces the system variation.

The next steps of the solution will optimize assuming a normal horsepower distribution, then optimize with a lognormal distribution. The purpose of using both distributions is to demonstrate the differences. In many cases, the error of assuming a normal distribution is trivial, and the required computations are less involved. For this problem, the error of assuming a normal distribution is large enough to be concerned about, and modeling a non-normal distribution is necessary.

[5]The initial solution has such a large standard deviation that the horsepower is significantly below zero in some cases. It is common for initial solutions to have unrealistic or even impossible outcomes. This is not concerning at the initial design point. These situations only become a concern when the final design is being considered.

7.4.4   Reservoir flow optimized with normal horsepower distribution

Before starting an optimization procedure, it is important to select the best target parameter. The reservoir flow example is trivial in that there is only a single response and there are no cost considerations. In later chapters, multiple responses will be considered as well as costs. Four commonly used targets for optimization are:

- output standard deviation,

- output C_{pk},

- output desirability, and

- total system cost.

When there is a single output, minimizing the output standard deviation and centering the bias corrected output mean will maximize robustness if the output distribution is symmetrical. When the output distribution is not symmetrical, maximizing robustness requires shifting the bias corrected output mean away from the center of the specifications. Maximizing C_{pk} allows robustness to be optimized by simultaneously changing the bias corrected output mean and the output standard deviation. Of course this requires the non-normal C_{pk} calculation.

Desirability (described in Chapter 6) allows multiple outputs to be combined into a single output. Desirability may or may not include a cost component. If cost is not included in the optimization routine it may be possible to obtain a very robust solution that is uncompetitive due to cost.

A total system cost can be found by combining all costs into a single output to be minimized. This technique is more complex, but yields a solution that optimizes the system rather than sub-optimizing a single part of the system. Components usually included in the system cost are:

- scrap and re-work cost for each output,

- processing cost for each input as a function of the input standard deviation,

- component cost for each input, and

- revenue as a function of each output (e.g., it may be possible to charge a higher price for a cellular phone if the battery life is increased).

Figure 7.12 shows the initial reservoir flow solution with C_{pk} computations. Since a normal distribution is being assumed for horsepower at this point, maximizing the C_{pk} will yield the same result as minimizing standard deviation.

Figure 7.13 shows the Solver setup to maximize the horsepower for the reservoir flow example.[6] The target cell is the cell that will be maximized. In this case, the target cell is B14 because cell B14 contains the calculation for horsepower C_{pk}. The

---

[6] This solution is contained on the accompanying website. The spreadsheet file name is ReservoirHorse-PowerNormal.xls. The Sage® file name is ReservoirFlowRobust.txt.

	A	B	C	D	E	F	G	H	I	J
1	Design parameter	Design nominal	Design constraint (lower)	Design constraint (upper)	Parameter standard deviation	h'	h''	Variance contribution	Standard deviation contribution	Bias
2	v	26.2	10	100	2	15.8	1.7	1003.0	31.7	3.44
3	d	4.7	0.5	10	0.1	-32.2	-15.8	10.4	3.2	-0.08
4	f	0.010	0.01	0.05	0.002	19271.2	0.0	1485.5	38.5	0.00
5	Δ	498.7	5	1000	3	-0.4	0.0	1.1	1.1	0.00
6	L	2000	2000	2000	1	0.1	0.0	0.0	0.1	0.00
7	h (theoretical) =	21.6								
8	h (actual) =	25.0								
9	h (STD) =	50.0								
10	Lower specification =	5								
11	Upper specification =	45								
12	C_{pk} (lower) =	0.133								
13	C_{pk} (upper) =	0.133								
14	C_{pk} =	0.133								

Figure 7.12   Reservoir flow example with C_{pk} for optimization.

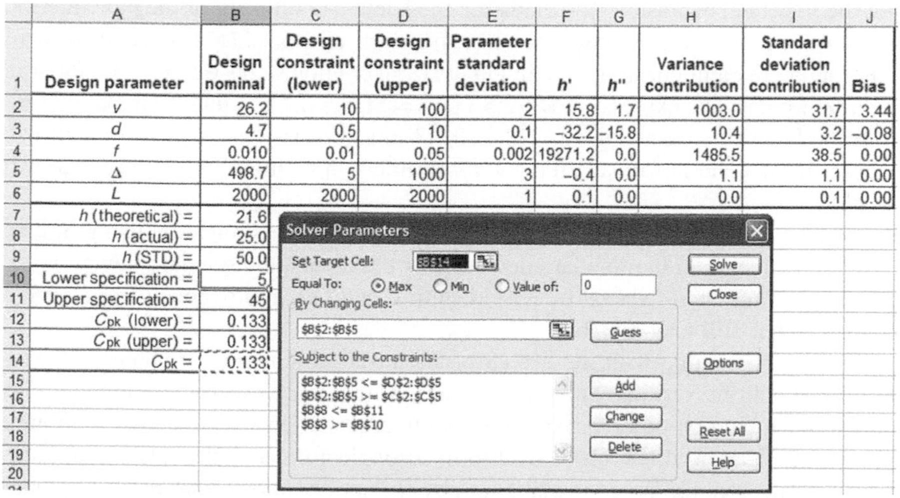

Figure 7.13   Horsepower C_{pk} maximization with Solver.[7]

horsepower C_{pk} will be maximized by changing the nominal values for the flow rate, the inside pipe diameter, the friction coefficient of the pipe interior, and the height difference between reservoir level and discharge level. The variables are located in cells B2 through B5 in the spreadsheet, and these cells are input into the *By Changing Cells* section of Solver. Solver has a *Subject to the Constraints* section to allow for any restrictions when optimizing. In this case, there are lower and upper constraints

---

[7]Depending on the version of Solver and the initial solution, Solver may fail to find the optimum solution. It may be useful to add a constraint forcing the mean near the center of the specifications. Optimization routines in Sage® and other mathematical routines may also fail to find the optimum in some cases. It is recommended that multiple starting points be used to avoid converging on a local optimum.

	A	B	C	D	E	F	G	H	I	J
1	Design parameter	Design nominal	Design constraint (lower)	Design constraint (upper)	Parameter standard deviation	h'	h''	Variance contribution	Standard deviation contribution	Bias
2	v	22.7	10	100	2	3.2	0.3	41.8	6.5	0.57
3	d	0.7	0.5	10	0.1	36.3	0.1	13.2	3.6	0.00
4	f	0.014	0.01	0.05	0.002	1808.0	0.0	13.1	3.6	0.00
5	Δ	5.0	5	1000	3	0.0	0.0	0.0	0.0	0.00
6	L	2000	2000	2000	1	0.0	0.0	0.0	0.0	0.00
7	h (theoretical) =	24.4								
8	h (actual) =	25.0								
9	h (STD) =	8.3								
10	Lower specification =	5								
11	Upper specification =	45								
12	C_{pk} (lower) =	0.808								
13	C_{pk} (upper) =	0.808								
14	C_{pk} =	0.808								

Figure 7.14   Spreadsheet solution for horsepower C_{pk} maximization with Solver.

for the flow rate, the inside pipe diameter, the friction coefficient of the pipe interior, and the height difference between reservoir level and discharge level. Instead of entering these eight constraints individually, Solver allows constraints to be input in column format. The expression $\$B\$2:\$B\$5 <= \$D\$2:\$D\$5$ is interpreted as cell B2 must be less than or equal to cell D2, cell B3 must be less than or equal to cell D3, etc. By using column format, all eight constraints can be input into Solver with two entries.

Figure 7.14 shows the spreadsheet solution for C_{pk} maximization using Solver. With no reduction in the standard deviation of the inputs, the standard deviation for horsepower has been reduced by 83% resulting in a C_{pk} improvement from 0.13 to 0.81, which is still unacceptable.

Before continuing with the design problem, the calculations contained in Figure 7.6 should be verified with simulation.[8] In addition to verifying computations at this point, the first-order variance approximation and the statistical distribution fit require verification. If the results from simulation do not match the variance approximation and there are no computational errors, then the more complicated second-order variance approximation must be used.

7.4.5   Reservoir flow optimized with normal horsepower distribution verification

Figure 7.15 shows the results of a 10 000-trial Monte Carlo simulation for the solution shown in Figure 7.14. The average horsepower from the simulation is 24.99, which is nearly identical to the solution found in the spreadsheet shown in Figure 7.6. The horsepower sample standard deviation is 8.7, which is slightly higher than the

---

[8] When creating optimization models in spreadsheets it is recommended to reduce the number of simulation trials before optimizing as every cell in every open spreadsheet is recalculated for every iteration of the optimization routine. With thousands of rows in a simulation this can take an excessive amount of time.

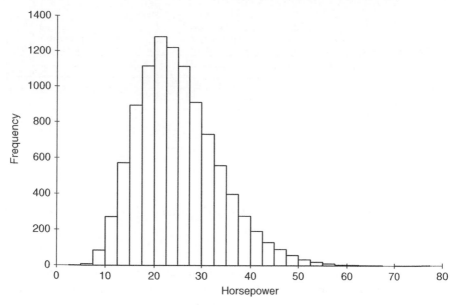

Figure 7.15   Monte Carlo simulation for verification of reservoir flow.

estimate obtained using the first-order approximation. It is expected that the first-order approximation will underestimate the true standard deviation, but if the error is reasonably small, the estimate can be used to optimize the system with final capability estimates coming from the simulation.

Figure 7.15 also shows the horsepower distribution is skewed right indicating a lognormal or Weibull distribution. It can be seen that there are no horsepower values below the lower specification of 5, while approximately 2.5% of the horsepower values exceed the upper specification of 45. Before continuing, a right skewed distribution should be used to model the horsepower distribution.

7.4.6   Reservoir flow horsepower variation sensitivity

Column I in the spreadsheet shown in Figure 7.14 shows the contribution of each input variable to the output standard deviation. The largest contributor to the horsepower standard deviation is v with a contribution of 6.5. The next largest contributors are d and f with contributions of 3.6 each. The inputs Δ and l have a contribution of approximately zero. To reduce the horsepower standard deviation, the input standard deviations must be reduced or constraints must be removed. Traditionally, an attempt is made to reduce the variation of all input parameters, however, this is an inefficient approach as reducing the standard deviation of Δ and l will not reduce the standard deviation of the horsepower. In fact, these parameters provide an opportunity for cost reduction.

	A	B	C	D	E	F	G	H	I	J
1	Design parameter	Design nominal	Design constraint (lower)	Design constraint (upper)	Parameter standard deviation	h'	h''	Variance contribution	Standard deviation contribution	Bias
2	v	22.7	10	100	2	3.2	0.3	41.8	6.5	0.57
3	d	0.7	0.5	10	0.1	36.3	0.1	13.2	3.6	0.00
4	f	0.014	0.01	0.05	0.002	1808.0	0.0	13.1	3.6	0.00
5	Δ	5.0	5	1000	12	0.0	0.0	0.0	0.1	0.00
6	L	2000	2000	2000	5	0.0	0.0	0.0	0.1	0.00
7	h (theoretical) =	24.4								
8	h (actual) =	25.0								
9	h (STD) =	8.3								
10	Lower specification =	5								
11	Upper specification =	45								
12	C_{pk} (lower) =	0.808								
13	C_{pk} (upper) =	0.808								
14	C_{pk} =	0.808								

Figure 7.16 Reservoir flow with increased standard deviation for parameters Δ and l.

Figure 7.16 shows the previous solution with the standard deviation of Δ increased from 3 to 12 and the standard deviation of l increased from one to five. Notice that the standard deviation of the horsepower did not change despite large increases in the standard deviation of these two input parameters. Input parameters with a small contribution to the output standard deviation should always be explored for cost reduction opportunities. Some examples of savings found by increasing the input standard deviation are:

- obtaining more parts from a tool before sharpening,

- reaming instead of honing,

- replacing a 1% resistor with a 5% resistor,

- using a less expensive material,

- using an older technology, or

- using a less expensive supplier.

Another opportunity for savings or improvement in output capability is constrained inputs. In some cases, it may be possible to alter a constraint, especially if the resulting savings or system performance outweighs the cost of changing the constraint. From Figure 7.14, it can be seen that the Δ and L are constrained. By changing the constraints for these parameters, it may be possible to reduce horsepower variation.

The horsepower standard deviation can also be decreased by reducing the standard deviation of the input parameters with the largest contribution to horsepower standard deviation. Unless it is significantly more expensive to reduce the standard deviation of flow rate than to reduce the standard deviation of diameter or pipe friction, the standard deviation of flow rate should be reduced first. If it is not possible to reduce the standard deviation of the flow rate, then the standard deviation for pipe inside diameter

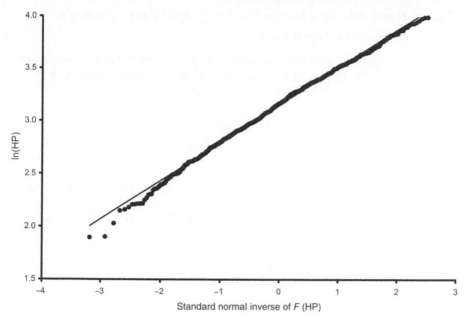

Figure 7.17   Lognormal probability plot for horsepower.

or pipe friction coefficient should be reduced. These reductions could possibly be made by purchasing a more expensive pipe or using a more expensive process. These options will be explored after a statistical distribution has been chosen to model horsepower.

7.4.7   Reservoir flow horsepower lognormal probability plot

When parameters are multiplied and divided, a skewed distribution often results, and it is also common for the distribution to be lognormal. A lognormal probability plot, shown in Figure 7.17, provides a graphical goodness-of-fit test for the horsepower output from the solution shown in Figure 7.14.

The points in Figure 7.17 form a reasonably straight line with no irregular pattern, indicating the lognormal distribution will provide a reasonable statistical model for horsepower.

It is common to find situations where there is no distribution that provides a reasonable statistical model for the system output. In these cases, the distribution with the closest fit should be used with the understanding that the optimized solution will have some inaccuracy. The final solution can be manipulated based on simulation results. When choosing a distribution in these cases, consider the specification limits of the output. If the output has a one-sided specification, it may be possible to find a distribution that has a poor overall fit but provides a reasonable prediction on a single tail.

7.4.8   Reservoir flow horsepower C_{pk} optimization using a lognormal distribution

From Figure 7.16, the estimated horsepower is 25, and the estimated variance (the sum of column H) is 68.1. The parameters of the lognormal distribution can be estimated as:

$$\sigma = \sqrt{2\ln(25) - \ln[(25)^2 - 68.1]} = 0.340$$

$$\mu = \ln(25) - \frac{0.347^2}{2} = 3.180$$

The probability of a value being less than 5 given a lognormal distribution with a log mean of 3.180 and a log standard deviation of 0.340 is $1.9(10^{-6})$. The probability of being greater than 45 for a value from the same distribution is 0.033. Thus, the C_{pk} is

$$C_{pk} = \frac{\Phi^{-1}(1 - 0.033)}{3} = 0.61$$

These computations can be added to the solution given in Figure 7.16 using the expressions shown in Figure 7.18.

After adding computations for C_{pk} computed with a lognormal horsepower distribution, the C_{pk} can be maximized as shown in Figure 7.19.

The solution displayed in Figure 7.19 shows a C_{pk} of 0.99. A portion of this improvement was gained by allowing the nominal horsepower to shift toward the lower specification. This result is due to the right skewness of the lognormal distribution.

The minimum allowable C_{pk} before a design should be released is 1.33, but a C_{pk} greater than 1.67 is preferred. Designing for a C_{pk} of 1.67 allows for degradation in performance due to uncertainties or changes after the design is released but before production begins. It is also important to remember that all variables may not be included in the equation being modeled. For example, factors such as temperature and water impurities may influence the horsepower but are not included in the model. Thus, the true C_{pk} is expected to be lower than the estimate from the model.

10	Lower specification =	5
11	Upper specification =	45
12	Log mean =	=LN(B8)-B13^2/2
13	Log standard deviation =	=(2*LN(B8)-LN(B8^2-B9^2))^0.5
14	Probability (HP<5) =	=LOGNORMDIST(B10,B12,B13)
15	Probability (HP>45 =	=1-LOGNORMDIST(B11,B12,B13)
16	C_{pk} =	=NORMSINV(1-(MAX(B14:B15)))/3

Figure 7.18 C_{pk} computations for lognormal horsepower.

	A	B	C	D	E	F	G	H	I	J
1	Design parameter	Design nominal	Design constraint (lower)	Design constraint (upper)	Parameter standard deviation	h'	h''	Variance contribution	Standard deviation contribution	Bias
2	v	20.8	10	100	2	2.3	0.2	20.5	4.5	0.44
3	d	0.6	0.5	10	0.1	25.4	0.0	6.5	2.5	0.00
4	f	0.012	0.01	0.05	0.002	1266.8	0.0	6.4	2.5	0.00
5	Δ	5.0	5.0	1000	3	0.0	0.0	0.0	0.0	0.00
6	L	2000	2000	2000	1	0.0	0.0	0.0	0.0	0.00
7	h (theoretical) =	15.6								
8	h (actual) =	16.1								
9	h (STD) =	5.8								
10	Lower specification =	5								
11	Upper specification =	45								
12	Log mean =	2.708								
13	Log standard deviation =	0.372								
14	Probability (HP<5) =	0.002								
15	Probability (HP>45) =	0.002								
16	C_{pk} =	0.985								

Figure 7.19 Maximization of C_{pk} computed with a lognormal horsepower distribution.[9]

Analyzing the results of the solution displayed in Figure 7.19, the design parameter v has the largest contribution to horsepower standard deviation at 4.5. Parameters d and f are tied with the next largest contribution at 2.5. The remaining parameters have a contribution near zero. Also, the parameters Δ and L are limited by design constraints.

The C_{pk} can be improved by relieving constraints from parameters Δ and L or reducing the standard deviation of the input parameters. For this example, relieving design constraints will not be explored.

If the costs of reducing the standard deviations of v, d, and f are in the same range, then the horsepower C_{pk} should be improved by reducing the standard deviation of v given it is the largest contributor to the horsepower standard deviation. If the cost of reducing the standard deviation of v is significantly higher than reducing the standard deviation of d or f then improvement in horsepower C_{pk} should be sought by reducing the standard deviation of d or f. In many cases, the standard deviation of multiple input parameters must be reduced to achieve the desired output C_{pk}.

Figure 7.20 shows the impact of reducing the standard deviation of v by 50%. The horsepower C_{pk} is increased from 0.99 to 1.32.

Further improvement to the solution found in Figure 7.20 is possible by optimizing the system again. Whenever any input parameter is changed, the optimum output may change. Figure 7.21 displays the results of running Solver on the solution shown in Figure 7.20. The horsepower C_{pk} is improved from 1.32 to 1.49.

Reducing the standard deviation of v by 50% did not achieve the desired horsepower C_{pk} of 1.67. Further improvement can be made by reducing the standard deviation of v even more, or by reducing the standard deviation of the next largest

---

[9]This solution is contained on the accompanying website. The file name is ReservoirHorsePowerLog-Normal.xls.

	A	B	C	D	E	F	G	H	I	J
1	Design parameter	Design nominal	Design constraint (lower)	Design constraint (upper)	Parameter standard deviation	h'	h''	Variance contribution	Standard deviation contribution	Bias
2	v	20.8	10	100	1	2.3	0.2	5.1	2.3	0.11
3	d	0.6	0.5	10	0.1	25.4	0.0	6.5	2.5	0.00
4	f	0.012	0.01	0.05	0.002	1266.8	0.0	6.4	2.5	0.00
5	Δ	5.0	5.0	1000	3	0.0	0.0	0.0	0.0	0.00
6	L	2000	2000	2000	1	0.0	0.0	0.0	0.0	0.00
7	h (theoretical) =	15.6								
8	h (actual) =	15.7								
9	h (STD) =	4.2								
10	Lower specification =	5								
11	Upper specification =	45								
12	Log mean =	2.719								
13	Log standard deviation =	0.274								
14	Probability (HP<5) =	0.000								
15	Probability (HP>45) =	0.000								
16	C_{pk} =	1.321								

Figure 7.20   Horsepower C_{pk} with v standard deviation reduced by 50%.

contributing parameters, d and f. Figure 7.22 shows the impact of reducing the standard deviation of d by 50% and optimizing. The horsepower C_{pk} is improved to 1.71, reaching the desired goal.

The design problem should not be considered complete because the horsepower C_{pk} goal has been achieved. The final step is to identify low contributing parameters for cost reductions. Figure 7.22 shows parameters Δ and L have an insignificant contribution to the horsepower standard deviation. It may be possible to reduce costs by increasing the standard deviation of Δ and L. Figure 7.23 shows the horsepower C_{pk} is unchanged when the standard deviation of Δ is increased from 3 to 12 and the standard deviation of L is increased from 1 to 5.

The final solution should always be verified with Monte Carlo simulation. Figure 7.24 shows a histogram from the results of a Monte Carlo simulation with

	A	B	C	D	E	F	G	H	I	J
1	Design parameter	Design nominal	Design constraint (lower)	Design constraint (upper)	Parameter standard deviation	h'	h''	Variance contribution	Standard deviation contribution	Bias
2	v	15.7	10	100	1	2.9	0.4	8.5	2.9	0.19
3	d	0.9	0.5	10	0.1	17.0	0.0	2.9	1.7	0.00
4	f	0.019	0.01	0.05	0.002	805.5	0.0	2.6	1.6	0.00
5	Δ	5.0	5.0	1000	3	0.0	0.0	0.0	0.0	0.00
6	L	2000	2000	2000	1	0.0	0.0	0.0	0.0	0.00
7	h (theoretical) =	15.3								
8	h (actual) =	15.5								
9	h (STD) =	3.7								
10	Lower specification =	5								
11	Upper specification =	45								
12	Log mean =	2.708								
13	Log standard deviation =	0.246								
14	Probability (HP<5) =	0.000								
15	Probability (HP>45) =	0.000								
16	C_{pk} =	1.491								

Figure 7.21   Horsepower C_{pk} optimized with v standard deviation reduced by 50%.

	A	B	C	D	E	F	G	H	I	J
1	Design parameter	Design nominal	Design constraint (lower)	Design constraint (upper)	Parameter standard deviation	h'	h''	Variance contribution	Standard deviation contribution	Bias
2	v	18.1	10	100	1	2.5	0.3	6.3	2.5	0.14
3	d	0.5	0.5	10	0.05	28.9	0.0	2.1	1.4	0.00
4	f	0.021	0.01	0.05	0.002	721.3	0.0	2.1	1.4	0.00
5	Δ	5.0	5.0	1000	3	0.0	0.0	0.0	0.0	0.00
6	L	2000	2000	2000	1	0.0	0.0	0.0	0.0	0.00
7	h (theoretical) =	15.2								
8	h (actual) =	15.3								
9	h (STD) =	3.2								
10	Lower specification =	5								
11	Upper specification =	45								
12	Log mean =	2.708								
13	Log standard deviation =	0.214								
14	P(HP<5) =	0.000								
15	P(HP>45) =	0.000								
16	C_{pk} =	1.714								

Figure 7.22   Horsepower C_{pk} optimized with d standard deviation reduced by 50%.

10 000 trials for the solution displayed in Figure 7.23. The average horsepower from the simulation provides a near exact match with the solution displayed in Figure 7.23, and the horsepower standard deviation is only slightly higher (3.28 vs. 3.24) than the solution displayed in Figure 7.23. Given the agreement in the two methods, there is no need for second-order estimate of horsepower standard deviation.

The histogram constructed from the Monte Carlo trials indicates agreement with a C_{pk} estimate of 1.71 as both tails are substantially buffered from the specification limits. The 1.71 estimate for horsepower C_{pk} is slightly high as the horsepower standard deviation estimate from the Monte Carlo simulation is 3.28. Before computing

	A	B	C	D	E	F	G	H	I	J
1	Design parameter	Design nominal	Design constraint (lower)	Design constraint (upper)	Parameter standard deviation	h'	h''	Variance contribution	Standard deviation contribution	Bias
2	v	18.1	10	100	1	2.5	0.3	6.3	2.5	0.14
3	d	0.5	0.5	10	0.05	28.9	0.0	2.1	1.4	0.00
4	f	0.021	0.01	0.05	0.002	721.3	0.0	2.1	1.4	0.00
5	Δ	5.0	5.0	1000	12	0.0	0.0	0.0	0.0	0.00
6	L	2000	2000	2000	5	0.0	0.0	0.0	0.0	0.00
7	h (theoretical) =	15.2								
8	h (actual) =	15.3								
9	h (STD) =	3.2								
10	Lower specification =	5								
11	Upper specification =	45								
12	Log mean =	2.708								
13	Log standard deviation =	0.214								
14	P(HP<5) =	0.000								
15	P(HP>45) =	0.000								
16	C_{pk} =	1.714								

Figure 7.23   Horsepower C_{pk} optimized with standard deviations for Δ and L increased.

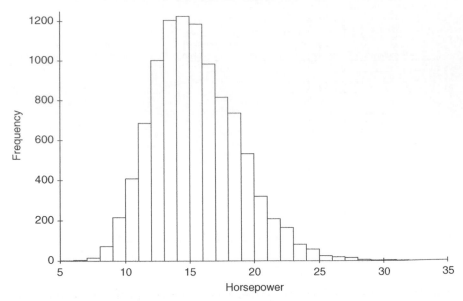

Figure 7.24   Monte Carlo simulation for horsepower.[10]

the C_{pk} with this standard deviation, the lognormal assumption for the horsepower distribution is verified with a lognormal probability plot as shown in Figure 7.25.

Using a mean of 15.35 and a standard deviation of 3.28, the parameters of the lognormal distribution can be estimated as:

$$\sigma = \sqrt{2\ln(15.35) - \ln\left[(15.35)^2 - (3.28)^2\right]} = 0.216$$

$$\mu = \ln(15.35) - \frac{0.216^2}{2} = 2.708$$

The probability of a value being less than five given a lognormal distribution with a log mean of 2.708 and a log standard deviation of 0.216 is $1.88(10^{-7})$. The probability of being greater than 45 for a value from the same distribution is $1.85(10^{-7})$. Thus, the C_{pk} is

$$C_{pk} = \frac{\Phi^{-1}\left[1 - 1.88(10^{-7})\right]}{3} = 1.69$$

The horsepower C_{pk} meets the goal of 1.67. At this point, a decision must be made regarding the quantity of variation in the system that is not represented in the

---

[10]The final solution to this problem is contained on the accompanying website. The file name is ReservoirHorsePowerLogNormalFinal.xls.

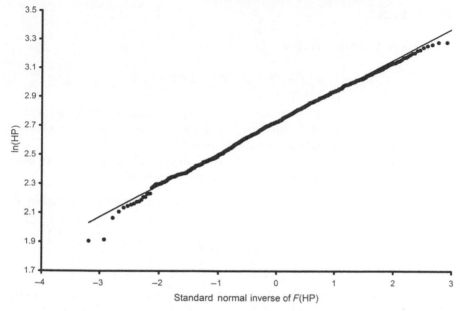

Figure 7.25   Horsepower lognormal probability plot.

horsepower equation. If that quantity of variation is small, and other sources of risk, such as the integrity of the estimated parameter standard deviations, are low, then the design should be considered complete. If these other risk factors are considerable, then more iterations should be taken to further improve horsepower C_{pk}.

7.5   Summary

The models for statistical bias and standard deviation can be used with optimization algorithms to minimize the standard deviation of the outputs without reducing the standard deviation of the inputs. These models are also used to compute process capability and determine the statistical behavior of the output while designs are still on paper. The ability to predict performance allows designs to be validated and allows the design engineer to determine the manufacturability of the design. A reasonable approximation to the optimum solution can be found by assuming normal distribution, or a skewed distribution, such as the lognormal or Weibull distribution, can be used to obtain a more exact solution.

In addition to optimizing the output, the model provides the variation contribution for each of the inputs. This allows any improvements to be targeted at the largest contributors and also allows parameters with low contributions to be used for cost reduction. The following chapters will extend these models to allow multiple outputs and total cost minimization.

Exercises

1   Consider the equation below.

$$X = 15Y^2 + 750Z + 240Y + 350$$

(a) Given $Z = 75$, what nominal value of Y yields a value of 125 000 for X?
(b) Given the standard deviation of Y is 6, and the standard deviation of Z is 3, what is the bias and standard deviation of X?
(c) Determine the nominal values of Y and Z which would minimize the variation of X and also achieve the required nominal X value of 125 000.

2   The amount of heat dissipated by a convective heat exchanger is governed by the equation

$$Q = \frac{(\Delta T)(A)}{C}$$

where ΔT is the temperature difference between the outside air and heat exchanger, A is the cross-sectional area of the heat exchanger, and C is the reciprocal of convective heat transfer coefficient.

The inputs are normally distributed with the following characteristics.

Variable	Lower Constraint	Upper Constrain	Standard Deviation
ΔT	10	50	0.02
A	5000	20 000	0.6
C	0.001	1	0.001

Identify nominal values for the inputs which would result in the least variation of output and also achieve a heat dissipation value of 100 000.

3   The outlet structure of a reservoir is used to regulate the flow of water out of the reservoir. The discharge at the outlet orifice needs to be maintained at a specified level. Excess variation in discharge would either cause flooding or may lead to water scarcity on the other side of the reservoir. A nominal discharge value of 40 000 m³/s is found out to ideal for utilizing the water effectively for agriculture. The outlet structure is a circular orifice and the design equation for discharge is given by

$$Q = \frac{CA}{\sqrt{2gH}}$$

where C is the orifice coefficient, A is the area of the orifice, and g is the acceleration due to gravity, and H is the water head in the reservoir.

Assume an orifice coefficient of 0.6 and acceleration due to gravity of 9.81 m/s^2. The diameter of the orifice is adjustable and due to its huge size it cannot be set accurately each time and so the area of the orifice varies and has a standard deviation of 3.5. The water head in the reservoir varies depending on the amount of rainfall received and from the historical data for the region we can compute a standard deviation of 15 for the head. Determine the nominal values of orifice area and head which would minimize the variation of discharge and also achieve the required nominal value of 40 000. The maximum nominal value for head has a constraint of 750 meters.

4　A machine tool manufacturing company supplies special purpose presses for various forging and stamping industries. One of the critical factors in a press is the amount of pressure applied, which is given by the transfer function

$$P = \frac{Ma}{A}$$

where M is the mass of the die, a is the acceleration of the die, and A is the area of the die.

(a) Given the constraints $M < 150\ 000$ N, $a < 50$ m/s^2, and $A < 1200$ m^2, determine the nominal values of input which would cause the least variation of pressure and also achieve a pressure rating of 7560 N/m^2.

(b) Assuming normal distribution for pressure, calculate the C_{pk} for the press, if the specifications for pressure are 7560 ± 500.

(c) Identify the distribution which best describes the distribution of pressure.

(d) Determine the C_{pk} for the press, using the distribution identified in (c).

5　What variable contributes to the majority of the output variance in Exercise 7.1? What is the output standard deviation if the standard deviation for this variable is reduced by 50%?

6　What variable contributes to the majority of the output variance in Exercise 7.2? What is the output standard deviation if the standard deviation for this variable is reduced by 50%?

7　What variable contributes to the majority of the output variance in Exercise 7.3? What is the output standard deviation if the standard deviation for this variable is reduced by 50%?

8　The equation which governs the power generated by the turbine is

$$P = 0.5DAV^3$$

where D is the air density, A is the rotor swept area, and V is the wind velocity.

The data collected for the past five years from various locations where the turbine is installed show that the wind velocity and air density follow a normal distribution. The variation in rotor swept area is due to the process capability of the manufacturer and follows a normal distribution with a mean of 500 and standard deviation of 2. The mean air density is 1.2 with a standard deviation of 0.1, and the mean wind velocity is 20 with a standard deviation of 2. Industry standards dictate that a turbine having a rotor swept area as given above should generate a power output of $2.4(10^6) \pm 4(10^5)$.

Design Constraints	Lower Limit	Upper Limit
Air density	0.8	1.4
Rotor swept area	350	1500
Wind velocity	10	60

(a) Which of the two inputs has the highest contribution towards the output standard deviation? What is the C_{pk} when the standard deviation of each these inputs is reduced by 50%?

(b) What specification allows a C_{pk} of 1.67?

8

Modeling system cost and multiple outputs

Chapter 7 presented methods for optimizing variation in design outputs. This chapter extends the optimization to consider additional factors such as component cost and manufacturing cost. This is an alternative to using the desirability index explained in Chapter 6. The methods presented in Chapter 7 are also extended to systems with multiple outputs, and offer techniques for large systems having hundreds or even thousands of inputs.

8.1   Optimizing for total system cost

Achieving a design that is insensitive to input variation is of no use if the design is too expensive to take to market. Consider the circuit example discussed in earlier chapters. If an attempt is made to minimize the sensitivity of power to voltage and resistance, the result will be a very large nominal voltage and resistance. The physics of the system yield lower power sensitivity as resistance increases.

Minimizing output sensitivity is considered a good design practice, but this objective is not independent. Insensitivity reduces scrap and re-work, which have associated costs. These costs must be balanced against the cost of the design. There is no value in reducing the scrap and re-work costs by 50% if the design costs are increased by more than 50%.

Probabilistic Design for Optimization and Robustness for Engineers, First Edition.
Bryan Dodson, Patrick C. Hammett and René Klerx.
© 2014 John Wiley & Sons, Ltd. Published 2014 by John Wiley & Sons, Ltd.
Companion website: http://www.wiley.com/go/robustness_for_engineers

	A	B	C	D	E	F	G	H
			Input standard			Output	Output standard	
1	Parameter	Nominal	deviation	P'	P"	variance	deviation	Bias
2	Voltage	12	0.9	6.67	0.56	36.00	6.00	0.23
3	Resistance	3.6	0.5	-11.11	6.17	30.86	5.56	0.77
4	Power (theoretical)	40.00						
5	Power (actual)	41.00			Cost model - per circuit			
6	Power (standard deviation)	8.18			Out of specification cost			$9.209
7	Lower specification	39.20			Voltage cost			$0.394
8	Upper specification	40.80			Resistance cost			$0.151
9	Log mean parameter	3.69			Total cost			$9.754
10	Log STD parameter	0.20						
11	P(power < 39.2)	45.16%			Process capability			
12	P(power > 40.8)	46.94%			Maximum out of specification			46.94%
13	P(out of specification)	92.09%			C_{pk}			0.03

Figure 8.1   Circuit power cost optimization initial solution.

Example 8.1[1]

The power in a circuit is

$$P = \frac{V^2}{R} \tag{8.1}$$

where V is the voltage and R is the resistance.

Given the following problem parameters, determine the nominal values for voltage and resistance that optimize the system cost.

- The power specification is 40 ± 0.8 watts.

- Each circuit that does not meet specifications causes a penalty of $10.

- The voltage cost is $0.25 + $0.012V$.

- The resistance cost is $0.15 + $0.0002R$.

- Voltage is normally distributed with a standard deviation of 0.9.

- Resistance is normally distributed with a standard deviation of 0.5.

Solution

Figure 8.1 displays an initial solution found by using 12 volts as the nominal voltage. Power variance is found with the first-order approximation. The percentage of circuits not meeting the specification is estimated with the lognormal distribution since power has been shown to follow a lognormal distribution in previous examples. This solution is unacceptable with over 90% of all circuits not meeting the specifications, resulting in a $9.754 cost per circuit.

---

[1] This example is contained on the accompanying website. The file name is CircuitCostOptimization.xls.

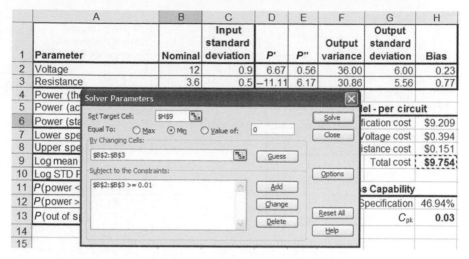

	A	B	C	D	E	F	G	H
			Input standard			Output	Output standard	
1	Parameter	Nominal	deviation	P'	P''	variance	deviation	Bias
2	Voltage	12	0.9	6.67	0.56	36.00	6.00	0.23
3	Resistance	3.6	0.5	-11.11	6.17	30.86	5.56	0.77
4	Power (th							
5	Power (ac						el - per circuit	
6	Power (sta						fication cost	$9.209
7	Lower spe						Voltage cost	$0.394
8	Upper spe						istance cost	$0.151
9	Log mean						Total cost	$9.754
10	Log STD F							
11	P(power <						s Capability	
12	P(power >						Specification	46.94%
13	P(out of s						C_{pk}	0.03
14								
15								

Solver Parameters

Set Target Cell: $H$9
Equal To: ○ Max ⦿ Min ○ Value of: 0
By Changing Cells:
$B$2:$B$3
Subject to the Constraints:
$B$2:$B$3 >= 0.01

[Solve] [Close] [Guess] [Options] [Add] [Change] [Reset All] [Delete] [Help]

Figure 8.2   Circuit power cost optimization Solver settings.

Figure 8.2 displays the Solver settings to minimize the system cost which consists of:

- out-of-specification cost,

- voltage cost, and

- resistance cost.

A constraint is required to force voltage and resistance to be positive. Since Solver does not have a *greater than* option, the constraint is *greater than or equal to* 0.01. The value 0.01 was chosen arbitrarily, any value above zero and sufficiently low not to constrain the solution could have been chosen.

Figure 8.3 gives the solution after minimizing total cost. By increasing voltage from 12 volts to 173.201 volts, the total cost per circuit is reduced from $9.75 to $3.18. This solution is still unacceptable, as the C_{pk} is 0.64.

Optimizing total cost rarely results in a solution with acceptable process capability. An acceptable C_{pk} results in a very low nonconformance rate, resulting in a negligible effect on total cost. To obtain an acceptable solution, a constraint must be added to force C_{pk} to be greater than 1.67. Optimization routines have difficulty with C_{pk} as a constraint because C_{pk} is computed from probability which is very low; the equivalent probability for a C_{pk} of 1.67 is $2.87(10^{-7})$.[2] This can be overcome by transforming C_{pk} into parts per million (ppm) and setting constrains against ppm. A C_{pk} constraint of 1.67 becomes a ppm constraint of 0.287. It may also be necessary to add constraints to force the nominal outputs between the specifications. In many

---

[2] The non-conformance probability is for the specification nearest to the mean.

	A	B	C	D	E	F	G	H
			Input standard			Output	Output standard	
1	Parameter	Nominal	deviation	P'	P"	variance	deviation	Bias
2	Voltage	173.201	0.9	0.46	0.00	0.17	0.42	0.00
3	Resistance	750.049	0.5	−0.05	0.00	0.00	0.03	0.00
4	Power (theoretical)	40.00						
5	Power (actual)	40.00			Cost model - per circuit			
6	Power (standard deviation)	0.42			Out of specification cost			$0.548
7	Lower specification	39.20			Voltage cost			$2.328
8	Upper specification	40.80			Resistance cost			$0.300
9	Log mean parameter	3.69			Total cost			$3.176
10	Log STD parameter	0.01						
11	P(power < 39.2)	2.70%			Process capability			
12	P(power > 40.8)	2.77%			Maximum out of specification			2.77%
13	P(out of specification)	5.48%			C_{pk}			0.64

Figure 8.3   Circuit power cost optimized for total cost.

cases, the constraints will have to be tighter than the specifications or optimization routines may be unable to locate the global optimum. Figure 8.4 displays the Solver settings to minimize power while maintaining a power C_{pk} greater than 1.67.

Figure 8.5 shows the solution for the circuit with minimum total cost and power C_{pk} greater than 1.67. The solution is validated with a 10 000-trial Monte Carlo simulation, which yields nearly identical results for the power average, standard deviation, and C_{pk}. The optimized solution calls for nominal voltage to be 461.933 and nominal resistance to be 5335.59. These values would not be used in the final design as it is unlikely these exact values would be available. Standard resistors are

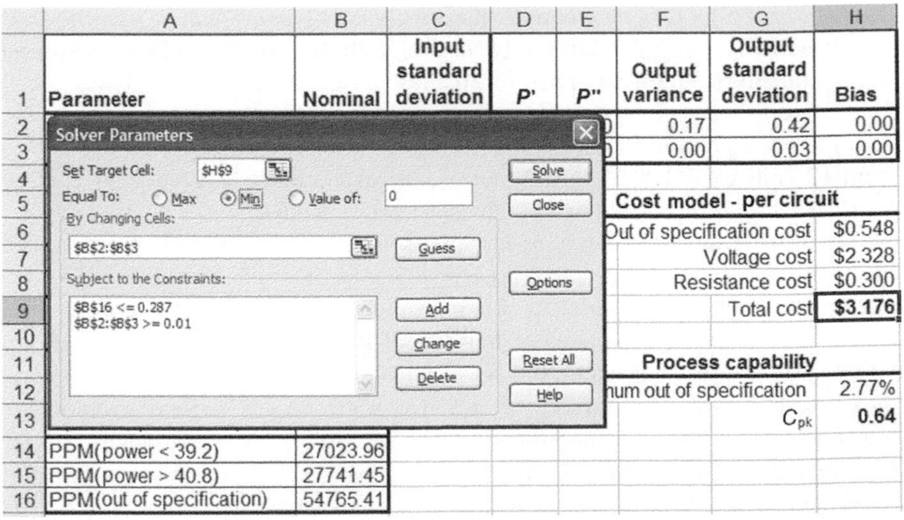

	A	B	C	D	E	F	G	H
			Input standard			Output	Output standard	
1	Parameter	Nominal	deviation	P'	P"	variance	deviation	Bias
2						0.17	0.42	0.00
3						0.00	0.03	0.00
4								
5						Cost model - per circuit		
6						Out of specification cost		$0.548
7						Voltage cost		$2.328
8						Resistance cost		$0.300
9						Total cost		$3.176
10								
11						Process capability		
12						um out of specification		2.77%
13						C_{pk}		0.64
14	PPM(power < 39.2)	27023.96						
15	PPM(power > 40.8)	27741.45						
16	PPM(out of specification)	54765.41						

Solver Parameters dialog box:
- Set Target Cell: $H$9
- Equal To: ○ Max ⊙ Min ○ Value of: 0
- By Changing Cells: $B$2:$B$3
- Subject to the Constraints:
  - $B$16 <= 0.287
  - $B$2:$B$3 >= 0.01
- Buttons: Solve, Close, Guess, Options, Add, Change, Delete, Reset All, Help

Figure 8.4   Solver settings for circuit power cost optimization including C_{pk} constraint.

	A	B	C	D	E	F	G	H
1	Parameter	Nominal	Input standard deviation	P'	P''	Output variance	Output standard deviation	Bias
2	Voltage	461.933	0.9	0.17	0.00	0.02	0.16	0.00
3	Resistance	5335.591	0.5	−0.01	0.00	0.00	0.00	0.00
4	Power (theoretical)	39.99						
5	Power (actual)	39.99			Cost model - per circuit			
6	Power (standard deviation)	0.156			Out of specification cost			$0.000
7	Lower specification	39.20			Voltage cost			$5.793
8	Upper specification	40.80			Resistance cost			$1.217
9	Log mean parameter	3.69			Total cost			**$7.010**
10	Log STD parameter	0.00						
11	$P(power < 39.2)$	0.00%			Process capability			
12	$P(power > 40.8)$	0.00%			Maximum out of specification			0.00%
13	$P(out of specification)$	0.00%			C_{pk}			**1.71**
14	PPM(power < 39.2)	0.143						
15	PPM(power > 40.8)	0.144						
16	PPPM(out of specification)	0.287						

Figure 8.5   Circuit power cost optimization with C_{pk} constraint.

available in 5100 ohms or 5600 ohms, and one of these values would likely be chosen, and a standard battery or power supply would also be used. This problem could also be expanded to include multiple resistors to enable a more precise resistance value to be obtained.

Including standard values as choices when optimizing is not recommended. Optimization routines have difficulty with step functions, and often fail when step functions are involved. It is best to start with a continuous function, and find a point close to the optimum. After an optimum region has been identified, the alternatives dictated by the solutions available can be explored (5100- versus 5600-ohm resistor).

Total cost models are useful for comparing alternatives. Decisions regarding cost versus performance can be made based on the total cost of the system rather than expert opinion.

Example 8.2[3]

What is the maximum voltage price increase that should be considered if the standard deviation of voltage is reduced from 0.9 to 0.5?

Solution

Figure 8.6 gives the optimum cost solution after reducing voltage standard deviation to 0.5 and optimizing. The solution is validated with a 10 000-trial Monte Carlo simulation, which yields nearly identical results for the power average, standard deviation, and C_{pk}. By reducing the voltage standard deviation, the nominal voltage

---

[3] This example is contained on the accompanying website. The file name is CircuitCostOptimization2.xls.

	A	B	C	D	E	F	G	H
1	Parameter	Nominal	Input standard deviation	P'	P''	Output variance	Output standard deviation	Bias
2	Voltage	257.329	0.5	0.31	0.00	0.02	0.16	0.00
3	Resistance	1655.774	0.5	−0.02	0.00	0.00	0.01	0.00
4	Power (theoretical)	39.99						
5	Power (actual)	39.99			Cost model - per circuit			
6	Power (standard deviation)	0.156			Out of specification cost			$0.000
7	Lower specification	39.20			Voltage cost			$3.338
8	Upper specification	40.80			Resistance cost			$0.481
9	Log mean parameter	3.69			Total cost			$3.819
10	Log STD parameter	0.00						
11	P(power < 39.2)	0.00%			Process capability			
12	P(power > 40.8)	0.00%			Maximum out of specification			0.00%
13	P(out of specification)	0.00%			C_{pk}			1.71
14	PPM(power < 39.2)	0.143						
15	PPM(power > 40.8)	0.144						
16	PPM(out of specification)	0.287						

Figure 8.6   Circuit power cost optimization with reduced voltage standard deviation.

and resistance can be reduced, resulting in a total cost of \$3.82. The optimum total cost with a voltage standard deviation of 0.9 is \$7.01, resulting in a savings of \$3.19.

The examples above could be extended to allow the input standard deviation to be a function of the nominal. For example, if a 10% resistor is used, and assuming $C_{pk} = 1.33$, then the standard deviation would be

$$\sigma = \frac{0.1R}{4} \tag{8.2}$$

The model can then also be extended to include a cost for nominal values and a cost for standard deviations. As stated earlier, it is not advised to have a step function for cost. A cost for a 10% resistor, a cost for a 5% resistor, etc., will often cause the optimization routines to converge on a sub-optimum solution. A continuous function that approximated the step costs can be used to allow the optimization routing to converge. The results can then be used as a starting point to compare the discrete solutions that are feasible.

8.2   Multiple outputs

Desirability was presented in Chapter 6 as a method for handling systems with more than one output. An alternative to using desirability is to use the total cost model. With each output included in the model, an additional cost term is added for the output not meeting the specifications, and a process capability constraint is added. Further extensions are possible, such as constraints, or input standard deviations being a function of other inputs, such as parameter nominal values.

8.2.1 Optimization

Whether using desirability or total cost, optimization routines have increasing difficulty finding a global optimum as the model becomes more complex. Below is a procedure to increase the efficiency of optimization routines.

1. Disable any C_{pk} calculations using non-normal distributions. If the probability of not meeting the specification is zero or one (software algorithms often return exactly zero or exactly one when the level of precision is exceeded), the calculation will fail causing the optimization routine to fail.

2. Determine input values that center all outputs between the specifications, or in the case of one-side specifications, place all outputs in a region that meets the specification.

3. Optimize for minimum ppm using the desired distribution. In many cases, a ppm less than 0.287 ($C_{pk} = 1.67$) may not be feasible. If the ppm is unacceptable, either:

   • remove constraints or

   • reduce the standard deviation for one or more inputs.

4. Optimize for minimum cost while achieving the desired ppm.

5. Enable any previously disabled C_{pk} calculations.

It is also recommended to optimize several times using different starting points. If there are local optimums, using multiple starting points will help prevent the selection of a local optimum.

8.2.2 Computing nonconformance

If any of a system's outputs are nonconforming, then the system is considered nonconforming, but the sum of the probability of nonconformance for each output does not result in the nonconformance probability for the system. Consider a system with two outputs having nonconforming probabilities of 75% and 65%. The sum results in 140%, which is obviously ridiculous. If the outputs are independent, the probability of nonconformance for the system is

$$P_S = 1 - \prod_{i=1}^{n} \left(1 - P_i\right) \tag{8.3}$$

Frequently, the outputs of a system are not independent. Computing the exact system nonconformance probability in this case becomes tedious. Fortunately, the error in assuming independence is low if the probability of nonconformance is low. The error of summing the probability of nonconformance to estimate the probability of system conformance is also low when the probability of nonconformance is low for the individual outputs. Figure 8.7 shows the error of summing the probability of

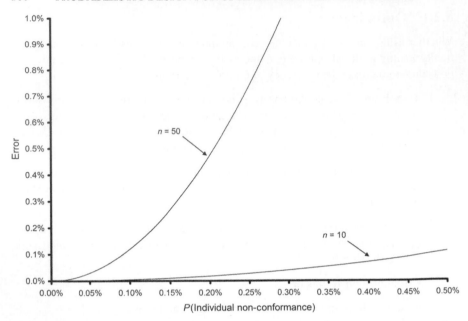

Figure 8.7   System nonconformance error when nonconformances are summed.

nonconformance versus using Equation 8.3. Since any reasonable solution will have a small probability of nonconformance for each output, many chose to approximate the system probability of nonconformance by summation.

If the number of outputs is large, consider computing a system C_{pk} from the nonconformance probability of the system. It is possible for every individual system output to have an acceptable C_{pk}, while the system has an unacceptable C_{pk}. Figure 8.8 shows the system C_{pk} versus the number of outputs in the system with each output having a C_{pk} of 1.67.

Example 8.3[4]

The work in joules and final pressure in pascals for a polytropic combustion process are

$$W = 1000 \left(\frac{P_1 V_1 - P_1 V_2 \left(\frac{V_1}{V_2} \right)^N}{N - 1} \right) \tag{8.4}$$

$$P_F = P_1 \left(\frac{V_1}{V_2} \right)^N \tag{8.5}$$

---

[4] This example is contained on the accompanying website. The file name is Polytropic_ Combustion.xls.

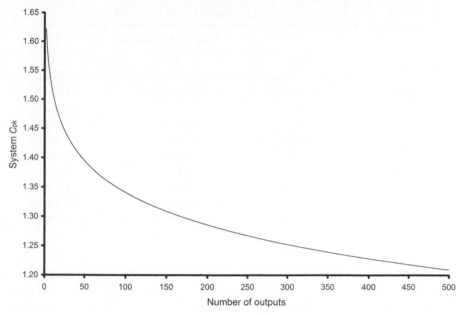

Figure 8.8   System C_{pk} as a function of the number of system outputs.

where P_1 is the initial pressure in pascals, V_1 is the initial volume in liters, V_2 is the final volume in liters, and N is the polytropic exponent which is dimensionless.

The work specification is 2400 J < W < 3100 J, and the final pressure specification is 2.3 Pa < P_F < 2.7 Pa. The parameter constraints and standard deviations are given in Table 8.1.

The cost per piece for not meeting the specifications is $500. The cost for the polytropic exponent is $4.2N. There is no cost associated with final volume, and the standard deviation of the final volume is equal to the standard deviation of the initial volume. The combined values of initial volume and initial pressure determine the mass of material used. The cost of the material is 0.8P_1 V_1$. Table 8.2 gives additional options for material cost and standard deviation combinations.

Determine the values of P_1, V_1, V_2, and N which minimize system cost and achieve a C_{pk} greater than 1.67 for W and P_2.

Table 8.1   Polytropic combustion parameter standard deviation.

Parameter	Lower Constraint	Upper Constraint	Standard Deviation
P_1	0.01	100	$0.02P_1$
V_1	0.01	100	$0.001V_1$
V_2	0.01	500	$0.001V_2$
N	1.5	5.0	0.01

Table 8.2   Cost and standard deviation options for Example 8.2.

Cost	Standard Deviation		
	P_1	V_1	V_2
$0.8P_1V_1$	$0.02P_1$	$0.001V_1$	$0.001V_2$
$0.4P_1V_1$	$0.04P_1$	$0.001V_1$	$0.001V_2$
$0.4P_1V_1$	$0.02P_1$	$0.002V_1$	$0.002V_2$
$0.8P_1V_1$	$0.01P_1$	$0.002V_1$	$0.002V_2$
$0.8P_1V_1$	$0.04P_1$	$0.0005V_1$	$0.0005V_2$
$1.2P_1V_1$	$0.01P_1$	$0.001V_1$	$0.001V_2$
$1.2P_1V_1$	$0.02P_1$	$0.0005V_1$	$0.0005V_2$
$1.6P_1V_1$	$0.01P_1$	$0.0005V_1$	$0.0005V_2$

Solution[5]

To estimate process capability and nonconformance cost, probability distributions must be determined for work and final pressure. Final pressure results from multiplication and division of input parameters, and is expected to be lognormally distributed. It is unclear what distribution should be used to model work. The subtraction term could result in a right- or left-skewed distribution depending on the magnitude. The normal distribution is used for an initial solution and will be verified with a Monte Carlo simulation. Figure 8.9 shows the initial solution resulting in a system C_{pk} of 1.2 and a system cost of $50.43.

Before optimizing for cost, it must be determined if the system is capable of reaching a C_{pk} greater than 1.67. Figure 8.10 displays the solution after maximizing for C_{pk}. The C_{pk} formula was disabled during optimization, and the goal of the optimization routine is to minimize the total ppm. If the percent nonconformance is very close to zero or one, the C_{pk} function will return an error causing the optimization routine to fail. Figure 8.10 shows the maximum achievable C_{pk} is 1.28, given the input constraints and standard deviations. The solution is constrained by the polytropic exponent (N) and final pressure. At this point, minimizing cost and accepting a C_{pk} of 1.26 results in a total system cost of $39.45, a 22% cost reduction over the initial solution.

The following options are available for improving the C_{pk}:

- change the lower constraint for N and

- reduce the standard deviation for P_1, V_1, or V_2.

If the lower constraint for N cannot be changed, the standard deviation of one of the inputs must be reduced. The preferred choice is P_1 since it has the largest

---

[5] This example uses C_{pk} computed from the probability of total system nonconformance. This is done only to demonstrate the method, as it is not required when the number of outputs is small.

	A	B	C	D	E	F	G	H	I	J
1					Work					
2			Constraints					Contribution		
3	Parameter	Nominal	Lower	Upper	Parameter standard deviation	W'	W''	Variance	Standard deviation	Bias
4	P_1	3.000	0.01	100	0.060	900.5	0.0	2919.0	54.03	0.00
5	V_1	15.727	0.01	100	0.016	−2485	−507	1527.0	39.08	−0.06
6	V_2	16.713	0.01	500	0.017	2499.7	−448.7	1745.4	41.78	−0.06
7	N	3.000	1.5	5.0	0.01	−80.5	3	0.6475	0.80	0.00
8	Work (theoretical)	2701								
9	Work (actual)	2701	2400	3100				6192.1	78.69	−0.1
10	PPM(W < 2400)	64.48								
11	PPM(W > 3100)	0.20								
12										
13					Final pressure (P_F)					
14			Constraints					Contribution		
15	Parameter	Nominal	Lower	Upper	Parameter standard deviation	P_F'	P_F''	Variance	Standard deviation	Bias
16	P_1	3.000	0.01	100	0.060	0.833	0.000	0.00250	0.0500	0.00
17	V_1	15.727	0.01	100	0.016	0.477	0.061	0.00006	0.0075	0.00
18	V_2	16.713	0.01	500	0.017	−0.449	0.107	0.00006	0.0075	0.00
19	N	3.000	1.5	5.0	0.010	−0.152	0.009	0.00000	0.0015	0.00
20	P_F (theoretical)	2.500								
21	P_F (actual)	2.500	2.3	2.7				0.00261	0.0511	0.00
22	Log mean	0.916								
23	Log STD	0.020								
24	PPM(P_F < 2.3)	24.39								
25	PPM(P_F > 2.7)	79.35								
26										
27	Process capability							Cost model - per piece		
28		PPM (total)	168.42					Out of specification cost		$0.08
29		C_{pk}	1.20					(P_1)(V_1) cost		$37.74
30								V_2 cost		$0.00
31								N cost		$12.60
32								Total cost		$50.43

Figure 8.9   Polytropic combustion initial solution.

contribution to final pressure standard deviation. The model can be used to determine the impact of reducing the input standard deviations. If the standard deviation for P_1 is reduced by 50%, the system C_{pk} increases to 1.87 without further optimization. Reducing the standard deviation of both V_1 and V_2 and optimizing has nearly no impact on the system C_{pk}. Since changing the standard deviation of V_1 and V_2 has a negligible impact on the system C_{pk}, increasing the standard deviations of these variables should be considered as a method of cost reduction.

Figure 8.11 gives a final solution obtained by reducing the standard deviation of P_1, increasing the standard deviation of V_1 and V_2 (the fourth row of Table 8.2), and optimizing. By designing and optimizing for total cost, the system C_{pk} is increased by 39% (from 1.20 to 1.67) and the total system cost is reduced by 83% (from $50.43 to $8.86). Modeling variation in addition to nominal performance allows nonlinear behavior to be exploited resulting in improved performance and reduced cost.

	Work								
		Constraints						Contribution	
Parameter	Nominal	Lower	Upper	Parameter standard deviation	W'	W''	Variance	Standard deviation	Bias
P_1	2.758	0.01	100	0.055	997.0	0.0	3025.1	55.00	0.00
V_1	15.005	0.01	100	0.015	−2483	−267	1388.4	37.26	−0.03
V_2	16.054	0.01	500	0.016	2492.3	−232.9	1600.9	40.01	−0.03
N	1.500	1.1	5.0	0.01	−92.4	4.2	0.8537	0.92	0.00
Work (theoretical)	2750								
Work (actual)	2750	2400	3100				6015.3	77.56	−0.1
PPM(W < 2400)	3.21								
PPM(W > 3100)	3.19								

	Final pressure (P_F)								
		Constraints						Contribution	
Parameter	Nominal	Lower	Upper	Parameter standard deviation	P_F'	P_F''	Variance	Standard deviation	Bias
P_1	2.758	0.01	100	0.055	0.904	0.000	0.00248	0.0498	0.00
V_1	15.005	0.01	100	0.015	0.249	0.008	0.00001	0.0037	0.00
V_2	16.054	0.01	500	0.016	−0.233	0.036	0.00001	0.0037	0.00
N	1.500	1.1	5.0	0.010	−0.168	0.011	0.00000	0.0017	0.00
P_F (theoretical)	2.492								
P_F (actual)	2.492	2.3	2.7				0.00252	0.0502	0.00
Log mean	0.913								
Log STD	0.020								
PPM(P_F < 2.3)	34.42								
PPM(P_F > 2.7)	33.47								

Process capability			Cost model - per piece	
PPM (total)	74.29		Out of specification cost	$0.04
C_{pk}	1.26		(P_1)(V_1) cost	$33.11
			V_2 cost	$0.00
			N cost	$6.30
			Total cost	$39.45

Figure 8.10   Polytropic combustion C_{pk} maximization.

This solution should not be accepted as correct without checking the results through Monte Carlo simulation. A 10 000-trial simulation shows nearly exact matches for the work and final pressure for both the averages and the standard deviations, and verifies the correct statistical distributions for both outputs. The simulation also tabulates zero nonconformances. With a requirement of less than one nonconformance per million, at least 10 million trials are required to validate this result. If 10 million trials are not feasible, histograms and probability plots could be used to ensure the correct distributions have been used to compute the nonconformance probabilities.

8.3   Large-scale systems

The polytropic combustion demonstrates how to model robustness for systems with multiple outputs, and while this example has only two outputs, it is easy to see

Parameter	Nominal	Constraints Lower	Upper	Parameter standard deviation	W'	W''	Variance	Standard deviation	Bias
Work									
P_1	13.425	0.01	100	0.134	204.8	0.0	756.2	27.50	0.00
V_1	0.238	0.01	100	0.000	3875	−48178	3.4	1.85	−0.01
V_2	0.733	0.01	500	0.001	2491.9	−5101.4	13.3	3.65	−0.01
N	1.500	1.5	5.0	0.01	−1400.0	997	196.0	14.00	0.05
Work (theoretical)	2750								
Work (actual)	2750	2400	3100				969.0	31.13	0.0
PPM(W < 2400)	0.00								
PPM(W > 3100)	0.00								

Parameter	Nominal	Constraints Lower	Upper	Parameter standard deviation	P_F'	P_F''	Variance	Standard deviation	Bias
Final pressure (P_F)									
P_1	13.425	0.01	100	0.134	0.186	0.000	0.00062	0.0249	0.00
V_1	0.238	0.01	100	0.000	15.677	32.876	0.00006	0.0075	0.00
V_2	0.733	0.01	500	0.001	−5.101	17.405	0.00006	0.0075	0.00
N	1.500	1.5	5.0	0.010	−2.798	3.141	0.00078	0.0280	0.00
P_F (theoretical)	2.492								
P_F (actual)	2.492	2.3	2.7				0.00152	0.0389	0.00
Log mean	0.913								
Log STD	0.016								
PPM(P_F < 2.3)	0.15								
PPM(P_F > 2.7)	0.14								

Process capability		Cost model - per piece	
PPM (total)	0.287	Out of specification cost	$0.00
C_{pk}	1.67	$(P_1)(V_1)$ cost	$2.56
		V_2 cost	$0.00
		N cost	$6.30
		Total cost	**$8.86**

Figure 8.11   Polytropic combustion final solution.

how this model could be expanded to include more outputs. The only difficulty when adding additional outputs is the tediousness of performing the calculations. It is not uncommon for systems to have hundreds of outputs and thousands of inputs. For example, an automobile audio unit may have over 200 outputs and over 10 000 inputs.

The difficulty of performing thousands of calculations can be overcome with high-level programming languages capable of performing symbolic mathematics. The accompanying website contains several of the examples in this text worked in the Sage® programming language.

Of course symbolic differentiation cannot be accomplished unless an equation representing the system output is available. The engineering of a system is often known, but there is no equation available. For example, when designing a bearing housing, finite element analysis (FEA) is used to compute many design outputs, but the FEA software is a black box and the engineer has no access to equations. In this case, derivatives can be taken numerically and an optimization algorithm can be used with the black box to optimize the system.

For robustness modeling, the inputs vary in a region approximately plus or minus three standard deviations from the nominal. Defining h as three standard deviations of the variable of interest, a numerical estimate of the first derivative of a function of several variables is

$$\frac{\partial f}{\partial x_1}(x_1, x_2, \ldots, x_n) \approx \frac{f(x_1 + h, x_2, x_3, \ldots, x_n) - f(x_1 - h, x_2, x_3, \ldots, x_n)}{2h}$$

(8.6)

A numerical estimate of the second derivative of a function of several variables is

$$\frac{\partial^2 f}{\partial x_1^2}(x_1, x_2, \ldots, x_n)$$
$$\approx \frac{f(x_1 + h, x_2, x_3, \ldots, x_n) - 2f(x_1, x_2, \ldots, x_n) + f(x_1 - h, x_2, x_3, \ldots, x_n)}{h^2}$$

(8.7)

Commercial engineering tools, such as FEA packages, allow inputs to be changed externally by exposing these inputs through objects and properties. This makes it possible to control the engineering tools from another programming language. This allows a robust model with thousands of inputs to be created and optimized. Figure 8.12 shows the variance contribution of the top 100 variables before and after optimization for a single output of an automobile body controller. The body controller electrical engineering outputs were modeled using PSpice®, and the robustness model including numerical derivatives and optimizations was performed by MATLAB® which called PSpice® when calculations from the electrical engineering model were needed. The controller consisted of more than a thousand components, but by utilizing the PSpice® model, robustness was analyzed resulting in a cost reduction of more than 25% while system C_{pk} was improved by more than 75%.

8.4   Summary

The techniques introduced in earlier chapters are easily expanded to include multiple variables and total system cost. If equations are not available, numerical approximations can be used from engineering tools to create a robustness model. The largest obstacle when models enlarge is the behavior of optimization algorithms. These algorithms will often fail to converge, and may find local optimums rather than global optimums. These difficulties can be overcome, by finding a feasible solution before beginning cost or robustness optimization, and ensuring there are no functions in your model that return errors.

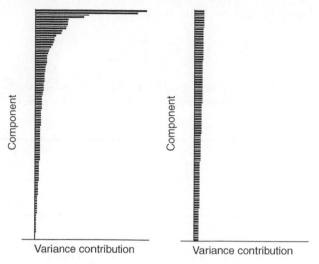

Variance contribution   Variance contribution

Figure 8.12   Body controller variable variance contributions before and after robustness.

The models presented in this chapter can be expanded further to provide total profit. For example, if battery life is an output, the selling price may be a function of battery life. There may be a lower specification for battery life, but as the battery life increases, the selling price increases along with profit.

Exercises

1 The deflection of the beam shown in the figure below is

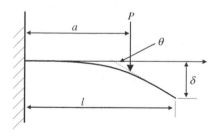

$$\delta = \frac{2Pa^2}{Ebh^3} (3l - a) \text{ meters}$$

where E is the Young's modulus (pascals), $I = bh^3/12$, b is the beam width (meters), and h is the beam height (meters).

The slope at the free end of the beam is

$$\theta = \frac{6Pa^2}{Ebh^3}$$

The specification for deflection is 0.25 ± 0.04 meters and the slope is required to be less than 0.07. The standard deviation and constraints for each of the parameters are:

Parameter	P	a	E	B	h	l
Standard deviation	0.02(P)	0.07	0.02(E)	0.002	0.002	0.01(l)
Lower constraint	5000	1.0	1.5×10^{11}	0.2	0.2	22
Upper constraint	5000	L	3.5×10^{11}	1.0	1.0	28

(a) What is the maximum C_{pk} given the constraints above?
(b) What parameter has the largest contribution to the system C_{pk}?
(c) Can system performance be improved by removing any constraints with the exception of P?
(d) If the standard deviation of a single variable can be reduced by 50%, what is the resulting system C_{pk}?

2   Given the beam in Exercise 8.1, and the following cost and standard deviation information, determine the nominal parameter values that minimize the system cost while achieving a system $C_{pk} > 1.67$. There is a nonconformance cost of 500.

Parameter	P	A	E	b	h	l
Standard deviation	0.01(P)	0.07	0.01(E)	0.002	0.002	0.005(l)
Cost	0	10a	1.5×10^{11}	0.2	0.2	22

3   A tank is filled with a volume of salt water (v) with a salt density of d. Fresh water flows in, there is perfect mixing, and salt water flows out. The flow rate in (f) is equal to the flow rate out. The requirements for Q are as follows:
• more than 90 grams of salt after 2 minutes
• between 70 and 90 grams of salt after 5 minutes
Standard deviations and constraints are provided below.

Parameter	d	v	f
Standard deviation	0.01	0.5	0.3
Lower constraint	1	1	0.3
Upper constraint	260	500	$v/5$

(a) What is the maximum C_{pk} given the constraints above?
(b) What parameter has the largest contribution to the system C_{pk}?
(c) Can system performance be improved by removing any constraints?
(d) If the standard deviation of a single variable can be reduced by 50%, what is the resulting system C_{pk}?

4 Given the beam in Exercise 8.3, and the following cost and standard deviation information, determine the nominal parameter values that minimize the system cost while achieving a system $C_{pk} > 1.67$. There is a nonconformance cost of 500.

Parameter	d	v	f
Standard deviation	0.01	0.5	0.3
Cost	$15(d)(v)$	$10(v)$	$5(f)$

5 In Exercise 8.4, how much does the cost increase if the standard deviation of f is doubled?

6 In Exercise 8.4, how much does the cost increase if the standard deviation of f is quadrupled?

9

Tolerance analysis

9.1 Introduction

In the design of new products, processes, or services, one of the final steps is the establishment of tolerances and specifications. These values establish the criteria for system acceptance, quantify the amount of allowable variation to meet customer and functional requirements, and ensure interchangeability during the assembly of components and sub-assemblies.

This interchangeability may be viewed from two perspectives. First, we have the interchangeability of the final product. Here, one must be able to randomly select any final product and have it fit and function as intended. In addition, appropriate tolerances ensure interchangeability of components and sub-assemblies.

In discussing the development of tolerances and specifications, we first provide operational definitions. We use the term *specification* to denote the actual permitted values from a nominal or target condition and the term *tolerance* to denote the relative allowable deviation from nominal. So, for a hole specification of 10 ± 0.2 mm, the nominal (or target) would equal 10, the plus or minus tolerance would equal 0.2, the tolerance width would equal 0.4 mm, and the lower and upper specification limits would be 9.8 and 10.2, respectively. Table 9.1 further summarizes some common terms used in tolerance analysis.

For establishing meaningful tolerances to ensure robust designs, tolerances should be viewed from a systems perspective, seeking a balance among meeting end-customer requirements, inherent process capability, cost, and component-to-sub-system relationships. One challenge in developing tolerances is conflicting perspectives. These conflicting perspectives may be summarized as the *designer's view* versus the *manufacturer's view* of tolerances. Under the *designer's view*, designers typically seek tolerances as tight as possible to ensure the product functions as intended. In

Probabilistic Design for Optimization and Robustness for Engineers, First Edition.
Bryan Dodson, Patrick C. Hammett and René Klerx.
© 2014 John Wiley & Sons, Ltd. Published 2014 by John Wiley & Sons, Ltd.
Companion website: http://www.wiley.com/go/robustness_for_engineers

Table 9.1   Tolerance analysis terminology.

Term	Code	Definition
Nominal (target)	N or T	Desired or specified value for a dimension
Lower specification limit	LSL	Actual lower or minimum specified value for a dimension
Upper specification limit	USL	Actual upper or maximum specified value for a dimension
Tolerance	t	Relative amount by which a dimension is permitted to vary from nominal
Equal bilateral	$\pm t$	Relative amount by which a dimension is permitted to vary from nominal in both directions
Unequal bilateral	$+t / -t$	Relative amount by which a dimension is permitted to vary from nominal in each direction
Unilateral tolerance	$+t$ or $-t$	Relative amount by which a dimension is permitted to vary from nominal in one direction
Tolerance width	W	Total tolerance width (USL − LSL)

contrast, under the *manufacturer's view*, wide tolerances are desired in hopes of lower operating costs and minimizing scrap and re-work.

Two classic ways to highlight these perspectives is to consider how each makes assumptions regarding the expected mean and standard deviation for a part produced in a new process. Designers tend to expect or assume that manufacturers will maintain the process mean at nominal as shown in Figure 9.1.

In contrast, manufacturers recognize that the mean may drift or shift as shown in Figures 9.2 and 9.3. A common cause of drift is tool wear. For a hole drilling operation, it is customary to set the nominal near the upper specification when a new drill bit is installed. As the drill bit wears, the hole becomes smaller, and when the hole size is near the lower specification, the drill bit is replaced. The frequency of drill bit changes may be used to control the overall process variation. When determining tolerances designers must not only consider drift during manufacturing, but also drift during a product's life. For example, electrolytic capacitors degrade over time, and the resistance in switches increase with age. Mean shifts have many causes, but changes in raw material and machine set-ups are examples.

Another deviation from controlled variation with the mean centered between the specifications is caused when the process output is sorted. When a production process is not capable either because the mean is shifted to one side of the specification (Figure 9.4), or the process is not capable (Figure 9.5), parts must be sorted.

A similar contrasting perspective occurs between manufacturers and their suppliers. Suppliers tend to push for tolerances as wide as possible to lower their costs.

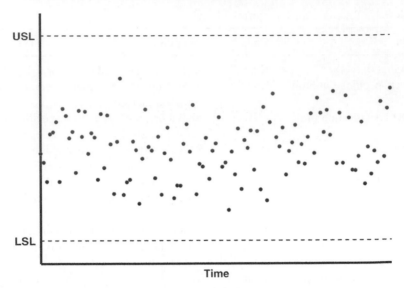

Figure 9.1   Controlled variation.

However, next process customers such as final assembly manufacturers often assign tighter tolerances than necessary to incorporate a safety factor to ensure product functional requirements are met. As Shewhart (1931) noted many years ago, which too often holds true today, design engineers expect that their specifications will be accepted by production engineers. Unlike production engineers, designers may not

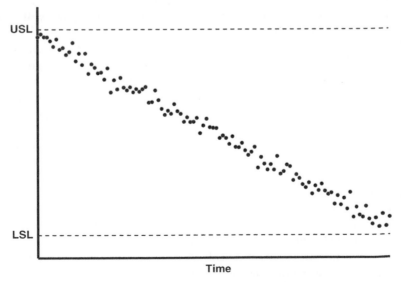

Figure 9.2   Variation with mean drift.

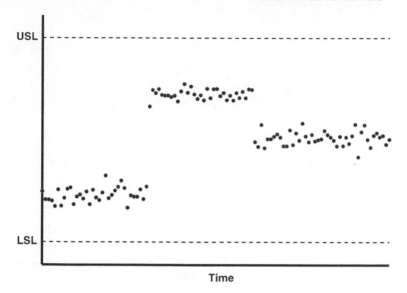

Figure 9.3   Variation with mean shift.

give serious consideration to production costs and feasibility. As a result, tolerance analysis among designers, suppliers, and manufacturing engineers often becomes subject to the behavioral characteristics of game theory strategies.

To help mitigate the gamesmanship in tolerance development, we discuss several approaches to tolerance analysis along with techniques to help identify appropriate

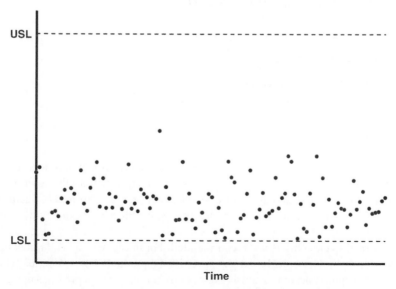

Figure 9.4   Variation for sorting with a mean shift.

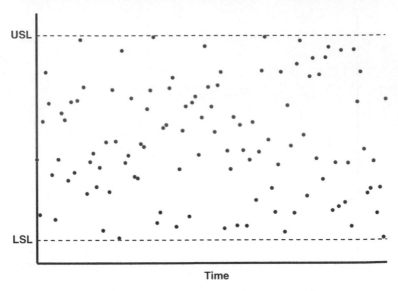

Figure 9.5   Variation after sorting with a un-capable process.

tolerances from a systems perspective. Specifically, we discuss several common approaches to tolerance analysis including worst-case and statistical stacking as well as tolerance design, which is an extension of statistical stacking methods and builds on the robust design methodology.

9.2   Tolerance analysis methods

The three most common methods for determining tolerances are:

- historical tolerancing,
- worst-case tolerancing, and
- statistical tolerancing.

9.2.1   Historical tolerancing

With historical tolerancing, tolerances are copied from historical products. This happens frequently when the engineering model is unknown. Initially, tolerances are set tight to reduce risk, and are later opened under pressure from manufacturing or because of costs. Over time, tolerances are opened to reduce costs and tightened when issues arise.

Historical tolerancing is not recommended, but in many cases, it is unavoidable because new products are based on an existing product. For example, when an automotive supplier designs a new brake system, it is based on a previous design. The new design is required because of, increased loading, decreased loading, increased

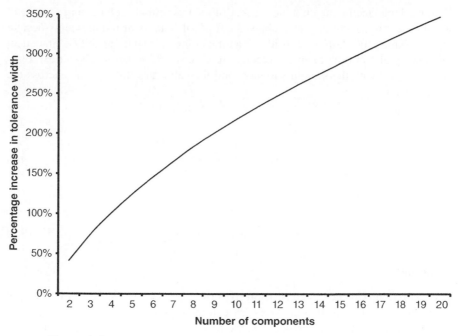

Figure 9.6   Excessive tolerance from worst-case tolerance stacking.

stiffness, or other requirement changes. In many cases, tolerances are scientifically determined for characteristics relating to the requirement change, but tolerances for the remaining characteristics are copied.

9.2.2   Worst-case tolerancing

Worst-case stacking is a simple and brute force method. As demonstrated with the brick wall in Chapter 3, worst-case stacking assumes the assembly joins extreme conditions for each of the components. The brick wall is a trivial example because wall height is the sum of the individual bricks, and the contribution to the system variation is equal for each component. As the number of components in a tolerance stack-up analysis increases, worst-case tolerancing will predict too much assembly variation (Figure 9.6). Liggett (1993) suggests a correction factor to be applied to worst-case tolerances.

9.2.3   Statistical tolerancing

There are two goals when determining tolerances:

1. ensure the design achieves all functional requirements despite the variation of the inputs, and

2. allow the manufacturing process to produce the design without any special controls.

Historical tolerancing may not meet either of the two primary tolerancing goals, and worst-case tolerancing ensures achievement of all functional requirements despite input variation, but fails to consider manufacturing cost and capability. Statistical tolerancing allows both primary tolerancing goals to be achieved. Statistical tolerancing begins with the system variance, and then allocates this variance across the system inputs.

Root sum square (RSS) is a statistical tolerancing method that applies to linear systems. If

$$y = x_1 + x_2 + \cdots + x_n \tag{9.1}$$

Then, the variance of y is

$$\sigma_y^2 = \sigma_{x_1}^2 + \sigma_{x_2}^2 + \cdots + \sigma_{x_n}^2 \tag{9.2}$$

Example 9.1[1]

The wall from Chapter 3 consists of 10 bricks stacked vertically with a height requirement of 100 ± 10 cm. Worst-case tolerances for the individual bricks is 10 ± 1 cm. Determine the brick tolerance using statistical tolerancing given a C_{pk} requirement of 1.67 for the wall.

Solution

The standard deviation requirement for the wall is

$$1.67 = \frac{110 - 100}{3\sigma_{wall}} \Rightarrow \sigma_{wall} = 2$$

Since the variance of a linear system is the sum of the individual variances, the standard deviation requirement for a brick is

$$\sigma_{wall} = 2 = \sqrt{10\sigma_{brick}^2} \Rightarrow \sigma_{brick} = 0.63$$

For inputs, a C_{pk} of 1.33 is required to avoid special manufacturing controls. With a C_{pk} of 1.33, and a brick standard deviation of 0.63, the upper brick specification is

$$1.33 = \frac{U - 1}{(3)(0.63)} \Rightarrow U = (1.33)(3)(0.63) + 1 = 3.51 \text{ cm}$$

---

[1] The example is included on the accompanying website. The file name is "Chapter_9_Wall_RSS.xlsx".

Since the brick nominal is 10, the brick tolerance is ± 2.51 cm, an increase of 151% from the worst-case tolerance of ± 1 cm. A Monte Carlo simulation confirms the wall C_{pk} is 1.67.

Equation 9.2 holds true for any combination of addition and subtraction; variances add for any linear combination of variables. Thus, RSS is commonly used for tolerance stacking situations, including gaps left from the differences of two stacks of components. While RSS is fine for simple, linear systems, the approach must be expanded to apply to more complex systems. The Taylor series approximations for variance and bias for complex systems can also be utilized for statistical tolerancing.

RSS is a simplification of the Taylor series approximations for variance and bias. The first derivative of Equation 9.1 with respect to x_1 is one, as is the first derivative with respect to each of the parameters. Using Equation 5.5, the variance of a sum is

$$\sigma^2 = \sum_{i=1}^{n} \left[(1)^2 \, \sigma_{x_i}^2 \right] = \sigma_{x_i}^2 + \sigma_{x_2}^2 + \cdots + \sigma_{x_n}^2 \tag{9.3}$$

Statistical tolerancing can be applied to any robust design model by multiplying the input standard deviation for each parameter by four if the input variation follows a normal distribution. Recall from Chapter 7 that if the mean is four standard deviations away from the closest specification, the C_{pk} equals 1.33. Also recall, that a C_{pk} of 1.33 is desired for inputs, but the target output C_{pk} is 1.67 to allow for any unknown influences, model assumptions, or other complications. In fact, in many cases, the input standard deviation is determined from the manufacturing specifications by applying a C_{pk} of 1.33. In other cases, the input standard deviation is measured directly. In either case, the tolerancing goal designing a product that can be manufactured without special controls is met.

For inputs that are not normally distributed, the specifications are determined by using the percentiles of the appropriate distribution, and will not be symmetrical. A C_{pk} of 1.33 is the equivalent of 31.67 ppm nonconformance to the nearest specification. The lower and upper specifications can be determined from the following expressions.

$$L = F^{-1}[31.67(10^{-6})] \tag{9.4}$$

$$U = F^{-1}[1 - 31.67(10^{-6})] \tag{9.5}$$

where $F^{-1}(x)$ is the inverse of the chosen cumulative distribution function.

The goal of ensuring the design achieves all functional requirements despite the variation of the inputs is achieved with the robust design model, however, the robust design model should always be verified with Monte Carlo simulation. If a first-order Taylor series expansion is used for the variance estimate, the output variance may be underestimated. The Monte Carlo simulation also verifies all assumptions regarding statistical distributions for both inputs and outputs.

Example 9.2[2]

Consider the polytropic combustion process from Example 8.3.

$$W = 1000 \left(\frac{P_1 V_1 - P_1 V_2 \left(\frac{V_1}{V_2} \right)^N}{N - 1} \right) \tag{9.6}$$

$$P_F = P_1 \left(\frac{V_1}{V_2} \right)^N \tag{9.7}$$

where P_1 is the initial pressure in pascals, V_1 is the initial volume in liters, V_2 is the final volume in liters, and N is the polytropic exponent, which is dimensionless.

Determine statistical tolerances for all input parameters.

Solution

Figure 9.7 shows the results after optimizing for system cost. Using the parameter input standard deviations from these results, the tolerances for the input parameters are computed in Table 9.2.

9.3 Tolerance allocation

In some cases, it is not possible to obtain information on manufacturing process capability. In these cases, tolerances can be equally allocated across the input parameters with unknown manufacturing process capability, by assigning equal output variance contributions to each input parameter.

Example 9.3[3]

The strain energy in a solid shaft is given in Equation 9.8. Given the information in Table 9.3, $0.46 < U < 0.54$, the process standard deviation for T is 11, and the process standard deviation for L is 0.53. Determine specifications for each input (allocate tolerances equally for G and r). What happens if the statistical tolerances were set to worst-case?

$$U = \frac{T^2 L}{G \pi r^4} \tag{9.8}$$

Solution

Figure 9.8 shows the results of a robust design model with specifications set at plus and minus four standard deviations from nominal. The variance contributions for G

---

[2] The example is included on the accompanying website. The file name is "Polytropic_Combustion_Tolerance.xls".

[3] This example is included on the accompanying website. The file name is "ShaftStrain.xlsx".

	A	B	C	D	E	F	G	H	I	J
1					Work					
2			Constraints					Contribution		
3	Parameter	Nominal	Lower	Upper	Parameter standard deviation	W'	W''	Variance	Standard deviation	Bias
4	P_1	13.425	0.01	100	0.134	204.8	0.0	756.2	27.50	0.00
5	V_1	0.238	0.01	100	0.00048	3875	−48178	3.4	1.85	−0.01
6	V_2	0.733	0.01	500	0.00147	2491.9	−5101.4	13.3	3.65	−0.01
7	N	1.500	1.5	5.0	0.01000	−1400.0	997	196.0	14.00	0.05
8	Work (theoretical)	2750								
9	Work (actual)	2750	2400	3100				969.0	31.13	0.0
10	PPM(W < 2400)	0.00								
11	PPM(W > 3100)	0.00								
12										
13					Final pressure (P_F)					
14			Constraints					Contribution		
15	Parameter	Nominal	Lower	Upper	Parameter standard deviation	P_F'	P_F''	Variance	Standard deviation	Bias
16	P_1	13.425	0.01	100	0.134	0.186	0.000	0.00062	0.0249	0.00
17	V_1	0.238	0.01	100	0.00048	15.677	32.876	0.00006	0.0075	0.00
18	V_2	0.733	0.01	500	0.00147	−5.101	17.405	0.00006	0.0075	0.00
19	N	1.500	1.5	5.0	0.01000	−2.798	3.141	0.00078	0.0280	0.00
20	P_F (theoretical)	2.492								
21	P_F (actual)	2.492	2.3	2.7				0.00152	0.0389	0.00
22	Log mean	0.913								
23	Log STD	0.016								
24	PPM(P_F < 2.3)	0.15								
25	PPM(P_F > 2.7)	0.14								

Figure 9.7   Polytropic combustion cost optimization results.

and r are forced to be equal by using Solver and setting the target cell to minimize the absolute value of the difference of the variance contribution of the two parameters. A constraint is added to Solver to force the ppm to be less than or equal to 0.287, which yields a $C_{pk} \geq 1.67$. A Monte Carlo simulation confirms the results.

Worst-case for U on the low side is found with T and L low, and G and r high. The opposite gives the worst-case for U on the high side. Using the results from Figure 9.8, the worst-case result for U on the low side is 0.444, and the worst-case result for U on the high side is 0.562.

9.4   Drift, shift, and sorting

In Example 9.3, the worst-case stack-up shows that when statistical tolerancing is used there is a possibility for the system to be nonconforming to specifications even though all input parameters are within specification. Statistical tolerancing assumes all inputs have controlled variation as shown in Figure 9.1. If the manufacturing process has drift, shift, or sorting as displayed in Figures 9.2, 9.3, and 9.4, statistical tolerancing may fail to produce outputs that adhere to specifications.

Table 9.2 Polytropic combustion statistical tolerancing results.

Parameter	Parameter Standard Deviation	Lower Specification	Upper Specification
P_1	0.134	$L = 13.425 - (4)(0.134) = 12.889$	$U = 13.425 + (4)(0.134) = 13.961$
V_1	0.00048	$L = 0.238 - (4)(0.00048) = 0.236$	$U = 0.238 + (4)(0.00048) = 0.240$
V_2	0.00147	$L = 0.733 - (4)(0.00147) = 0.727$	$U = 0.733 + (4)(0.00147) = 0.739$
N	0.01000	$L = 1.5 - (4)(0.01000) = 1.460$	$U = 1.5 - (4)(0.01000) = 1.540$

Table 9.3   Strain energy data.

Parameter	Nominal
T	2000
L	100
G	994 718.4
r	4

When faced with parameters that drift, the robustness model should include the input standard deviation in a narrow region (the beginning or the end of the process), and the model should be optimized for both conditions. There should be a statistical analysis when the drifting parameter is low, and when the drifting parameter is high. The results of these two optimizations can be used to determine tolerances.

In some cases, usage conditions vary. For example, an automobile's radio must function at $-20°C$ and at $120°C$. In this case, the temperature variation should be removed from the model, and the statistical analysis should be done at the temperature extremes to determine tolerances for the remaining parameters.

Statistical tolerancing is only effective when there is a small probability of tolerances lining up in worst-case conditions. When processes have mean shifts, this increases the chance of parameters lining up at worst-case conditions. If the parameter that has mean shifts is not a significant parameter, then the mean shift is not of concern. Recall the work output from the polytropic work in Example 8.3. For the final solution, 78% of the work variance is the result of variation of P_1, and that 98% of the work variance is the result of variation of P_1 and N. If P_1 or N have mean

	A	B	C	D	E	F	G	H	I	J
1					Strain energy due to torsion (*U*)					
2							Contribution		Specifications	
3	Parameter	Nominal	Parameter standard deviation	*U'*	*U"*	Variance	Standard deviation	Bias	Lower	Upper
4	*T*	2 000	11	0.000	0.000	0.00003	0.0055	0.00	1 956	2044
5	*L*	100	0.53	0.005	0.000	0.00001	0.0026	0.00	97.88	102.12
6	*G*	994 718.4	6559.72	0.000	0.000	0.00001	0.0033	0.00	968 480	1 020 957
7	*r*	4.00	0.00660	−0.500	0.625	0.00001	0.0033	0.00	3.974	4.026
8	*U* (theoretical)	0.500								
9	*U* (actual)	0.500				0.00006	0.0077	0.00		
10	Log mean	−0.693								
11	Log STD	0.015			Absolute value of cell F6 - cell F7		8E-09			
12	PPM(*U* < 0.46)	0.029								
13	PPM(*U* > 0.54)	0.272								
14										
15	Process capability									
16	PPM (Max)	0.272								
17	C_{pk}	**1.67**								

Figure 9.8   Strain energy in a shaft optimization and tolerances.

shift, then this must be accounted for in the statistical model. Mean shifts for other parameters may be ignored. Monte Carlo simulation is an accurate way to access the impact of the shift.

Sorting is similar to mean shifts, in that the chance of parameters lining up at worst-case conditions is increased. If the parameter that has mean shifts is not a significant parameter, then the sorting is not of concern. Sorting of significant parameters should be prohibited, and manufacturing should be notified of this by signifying critical to quality parameters.

9.5   Non-normal inputs

The central limit theorem states that the distribution of the sum or average tends to normal as the sample size increases regardless of the distribution of the inputs. With statistical tolerancing, the distribution of the output also becomes less sensitive to the distribution of the inputs as the sample size of the inputs increases, as long as the output variance is not dominated by a small number of variables. The Taylor series estimates for system variance and bias should always be confirmed by a Monte Carlo simulation.

9.6   Summary

Many factors affect the assignment of tolerances. In this chapter, we discussed different perspectives and approaches one may take to arrive at appropriate tolerances. Specifically, we discuss two classic methods: worst-case tolerancing and statistical tolerancing. The main benefit of statistical tolerancing with simulation as confirmation is a reduction in cost by increasing tolerance widths. Additional benefits are the visualization of results and the ability to explore what if scenarios by changing the assumptions of the model. Even with a comprehensive effort to develop a predictive tolerance model, identifying all of the factors and establishing appropriate estimates for their performance is a complex task. Tolerances must account for mean drift, shift, and sorting out of control processes.

Exercises

1   A supplier to the automobile industry produces a chassis frame for various luxury vehicles. The chassis requires an organic acrylic coating to protect it from corrosion during the useful life of the vehicle. The coating thickness is one of the key output characteristics. Suppose a new chassis is being developed and the design team wants to establish appropriate specification limits for the coating thickness so that the coating process will be able to achieve a $C_{pk} = 1.33$. By monitoring the current coating process, the standard deviation is estimated to be 0.021 mm. The desired target for the coating thickness is 0.190 mm.

(a) Assuming the process mean is at nominal, determine how large a tolerance is needed to achieve a $C_{pk} = 1.33$?

(b) Suppose changes in company policy requires all processes to achieve a C_{pk} of 1.67. The design team does not agree to widen the tolerances established previously in (a) as they believe it would affect the coating quality. By how much would the standard deviation of the coating process need to be reduced from current levels to achieve the new process capability goal?

2 A leading hotel chain is evaluating its check in process. Long check-in times causes customer dissatisfaction while hurrying up with the process causes lots of human errors. After a detailed time study, the service team has identified that it may take a minimum of 50 seconds to check in a customer and a maximum time of 100 seconds. The current average check-in time is 90 seconds with a standard deviation of 2.5 seconds. Does the current process meet the hotel's process capability requirement of 1.33?

3 A product design consists of five components in a linear stack. Suppose four of the component dimensions have specifications of 4.0 ± 0.1 mm, 3.0 ± 0.1 mm, 6.0 ± 0.1 mm, and 2.0 ± 0.1 mm, respectively. If each of these components follows a normal distribution, determine the specification needed for the fifth component given a final specification of 20.00 ± 0.25 mm. The required C_{pk} for the product is 1.67.

4 A fabrication company receives the following drawing from its customer. The assembly consists of four sheet metal parts where the gap (Y) indicated is a key characteristic of the assembly for the quality of the final product. The fabricator welds the four parts to make the assembly. Suppose data from the processes for each component follow a normal distribution with a standard deviation of 0.05 mm.

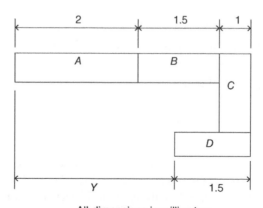

All dimensions in millimeters

(a) Given the process standard deviation, what tolerance would you propose for the component parts A, B, C, and D to achieve a C_{pk} goal of 1.33?

(b) Using the tolerances obtained in (a), employ statistical tolerancing to recommend a nominal dimension and a tolerance for the resultant gap.

5   One of the sub-assemblies of an industrial hoist is shown below. The gap A–B is a critical dimension and the manufacturer wants to identify a feasible assembly tolerance.

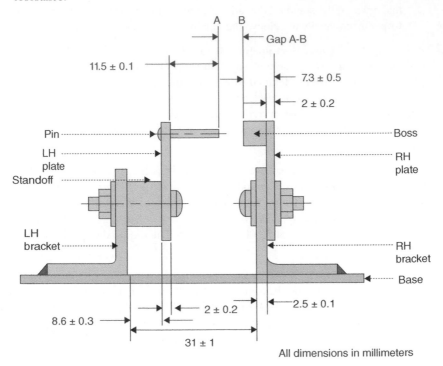

All dimensions in millimeters

P = Distance between RH and LH bracket = 31 ± 1
Q = RH bracket web thickness = 2.5 ± 0.1
R = RH plate thickness = 2 ± 0.2
S = Pin length = 11.5 ± 0.1
T = LH plate thickness = 2 ± 0.2
U = Standoff thickness = 8.6 ± 0.3
V = Thickness of RH plate and boss = 7.3 ± 0.5

(a) Given the above component information, use worst-case tolerancing to recommend a nominal dimension and assembly tolerance.

(b) Given the above component information, use statistical tolerancing to recommend a nominal dimension and assembly tolerance.

6   Use statistical tolerancing to determine tolerances for angle and velocity for the projectile example from Chapter 5. The required C_{pk} for distance is 1.67.

7   Use statistical tolerancing to determine tolerances for the parameters of the reservoir flow example from Chapter 7. The required C_{pk} for distance is 1.67.

10

Empirical model development

Chapter 8 discussed the methods for using numerical methods to optimize systems designed using black box software such as finite element analysis. In some cases, there is no engineering model available, whether in a black box or equations. In these cases, parameter values are determined from engineering standards, experience, or trial and error. This chapter presents a method for creating empirical models to estimate system outputs. There are many methods for doing this, but the method presented here emphasizes efficient experimentation. If the cost or time to perform experiments is not prohibitive, this method can be altered.

10.1   Screening

The first step when developing an empirical model is determining which parameters (or factors) to include in the model. A two-level fractional factorial experiment efficiently identifies relevant factors by assuming linearity.

A full factorial experiment requires L^F trials, where L is the number of levels and F is the number of factors. An advantage of using a full factorial experiment is that all components of the model can be estimated. The disadvantage of a full factorial is the number of experimental trials required. A two-level full factorial experiment with seven factors requires 128 trials. A two-level full factorial experiment with five factors requires 32 trials. Table 10.1 shows the number of experimental trials required to estimate the components of a two-level, five-factor experiment.

A fractional factorial reduces the number of experimental trials by confounding the components in the model. For example, a two-level, five-factor, $1/4$ fractional factorial experiment requires $1/4$ the number of trials, but does not give a clean estimate of the model components. The main effects are confounded with the interactions.

Probabilistic Design for Optimization and Robustness for Engineers, First Edition.
Bryan Dodson, Patrick C. Hammett and René Klerx.
© 2014 John Wiley & Sons, Ltd. Published 2014 by John Wiley & Sons, Ltd.
Companion website: http://www.wiley.com/go/robustness_for_engineers

Table 10.1   Components of a five-factor factorial with two levels.

Component	Number of Required Trials
Overall mean	1
Main effects	5
Two-way interactions	10
Three-way interactions	10
Four-way interactions	5
Five-way interaction	1

It is possible to begin with a fractional factorial and run additional trials from the full factorial to resolve the confounded components. This sequential approach provides estimates of all model components with a minimum of experimental trials. This methodology will be demonstrated with a five-factor experiment.

Table 10.2 shows the 32 experimental trials required for a two-level, five-factor full factorial design.[1] To provide degrees of freedom for experimental error, some of the 32 trials must be replicated.

An alternative to using a full factorial is to begin with a fractional factorial to reduce the number of required trials. Using the $1/4$ factorial design shown in Table 10.3 requires only eight trials.

Reducing the number of trials by 75% has a drawback. The full factorial shown in Table 10.2 allows estimates of the effects, for all factors, and all interactions among the factors. When the $1/4$ fractional factorial shown in Table 10.3 is used, the effects for the five factors cannot be separated from the interactions between the factors. The confounding structure for the $1/4$ fractional factorial is given in Table 10.4.

In the $1/4$ fractional factorial for five factors, the main effects are confounded with two-, three-, four-, and five-way interactions. If statistical analysis indicates that factor B is significant, it could be that the two-way interaction between factors A and D; the three-way interaction between factors C, D, and E; or the four-way interaction between factors A, B, C, and E is significant and factor B is not.

Example 10.1[2]

To illustrate the process, an empirical model will be developed for the noise in an assembly used in an automobile. For confidentiality purposes, the factors will only be identified with letters. Table 10.5 shows the results from a $1/4$ fractional factorial experiment with two replications. The replications are necessary to provide degrees of freedom to estimate experimental error, allowing a statistical analysis. There are

---

[1] Factorial experimental designs with confounding information for up to 255 factors are contained on the accompanying website. The file name is Factorial_Designs.xls.

[2] This example is contained on the accompanying website. The file name is AutomotiveNoise.xls.

Table 10.2   Five-factor full factorial with two levels.

Row	A	B	C	D	E
			Factor		
1	−1	−1	−1	−1	−1
2	1	−1	−1	−1	−1
3	−1	1	−1	−1	−1
4	1	1	−1	−1	−1
5	−1	−1	1	−1	−1
6	1	−1	1	−1	−1
7	−1	1	1	−1	−1
8	1	1	1	−1	−1
9	−1	−1	−1	1	−1
10	1	−1	−1	1	−1
11	−1	1	−1	1	−1
12	1	1	−1	1	−1
13	−1	−1	1	1	−1
14	1	−1	1	1	−1
15	−1	1	1	1	−1
16	1	1	1	1	−1
17	−1	−1	−1	−1	1
18	1	−1	−1	−1	1
19	−1	1	−1	−1	1
20	1	1	−1	−1	1
21	−1	−1	1	−1	1
22	1	−1	1	−1	1
23	−1	1	1	−1	1
24	1	1	1	−1	1
25	−1	−1	−1	1	1
26	1	−1	−1	1	1
27	−1	1	−1	1	1
28	1	1	−1	1	1
29	−1	−1	1	1	1
30	1	−1	1	1	1
31	−1	1	1	1	1
32	1	1	1	1	1

five factors in the experiment, but all seven factors from the factorial design are used when performing calculations to prevent the two unused columns from being used to estimate experimental error. The two unused columns will be labeled U 1 and U 2. The second and fifth trials were randomly selected for replication to provide degrees of freedom for experimental error.

Table 10.3   Five-factor $\frac{1}{4}$ fractional factorial with two levels.

		Factor		
A	B	C	D	E
1	−1	−1	−1	−1
−1	1	1	−1	−1
1	1	−1	1	−1
−1	−1	1	1	−1
−1	1	−1	−1	1
1	−1	1	−1	1
−1	−1	−1	1	1
1	1	1	1	1

Table 10.4   Confounding for
$\frac{1}{4}$ factorial five-factor experiment.

Overall Mean + ABD + ACE + BCDE
A + BD + CE + ABCDE
B + AD + CDE + ABCE
C + AE + BDE + ABCD
D + AB + BCE + ACDE
E + AC + BCD + ABDE
BC + DE + ABE + ACD
BE + CD + ABC + ADE

Table 10.5   Five-factor $\frac{1}{4}$ fractional factorial with responses.

Row	A	B	C	D	E	U 1	U 2	Response
1	18.33	0.004	0.004	350	54.1	1	−1	66.5
2	18.33	0.004	0.007	350	35.6	−1	1	65.8
3	18.33	0.007	0.004	175	54.1	−1	1	51.9
4	18.33	0.007	0.007	175	35.6	1	−1	52.2
5	24.81	0.004	0.004	175	35.6	1	1	71.6
6	24.81	0.004	0.007	175	54.1	−1	−1	70.5
7	24.81	0.007	0.004	350	35.6	−1	−1	50.4
8	24.81	0.007	0.007	350	54.1	1	1	51.8
9	18.33	0.004	0.007	350	35.6	−1	1	66.4
10	24.81	0.004	0.004	175	35.6	1	1	71.8

Table 10.6   Statistical analysis of the $1/4$ fractional factorial using coded units.

Factor	Coefficient	Standard Error	p Value[3]	95% Confidence Interval	
				Lower	Upper
Intercept	60.14	0.105	0.0%	59.69	60.59
A	0.96	0.105	1.2%	0.51	1.41
B	−8.56	0.105	0.0%	−9.01	−8.11
C	0.01	0.105	91.6%	−0.44	0.46
D	−1.44	0.105	0.5%	−1.89	−0.99
E	0.04	0.105	75.4%	−0.41	0.49
U1	0.41	0.105	5.9%	−0.04	0.86
U2	0.24	0.105	15.1%	−0.21	0.69

Statistical analysis of the data in Table 10.5 yields factors A, B, and D as statistically significant at a significance level of 5%. The regression is conducted with coded units, −1 for the low value and 1 for the high value. Using coded units maintains balance in the experimental design, and having all factors on the same scale allows coefficients to be compared. This is useful when statistical significance is marginal, and in some cases, a statistically significant factor may be removed from the model because of a low contribution (low coefficient), to obtain a simpler model. This is shown in Table 10.6.

At this point, it cannot be concluded that factors A, B, and D are statistically significant because of the confounding in the experimental design. Recall from Table 10.4, the confounding for factors A, B, and D is:

A + BD + CE + ABCDE
B + AD + CDE + ABCE
D + AB + BCE + ACDE

The next step of the analysis is to separate the effects of the confounded components. To reduce the number of experimental trials, the lower-level interactions will be investigated first, and higher-level interactions will be pursued if the main effects and lower-level interactions are not statistically significant. Taking this approach, only the main effects and the two-way interactions will be pursued with the next iteration of experiments.

There are four, two-way interactions that are confounded with the main effects. Three of these two-way interactions, BD, AD, and AB, consist of combinations of the main effects that may be statistically significant. Because of this, the interaction

---

[3]The p value is used for determining statistical significance. As a rule of thumb, if the p value is less than 5% the factor is statistically significant; and if the p value is greater than 5% it is not statistically significant.

Table 10.7   Three-factor full factorial.

A	B	D
−1	−1	−1
−1	−1	1
−1	1	−1
−1	1	1
1	−1	−1
1	−1	1
1	1	−1
1	1	1

between factors C and E will be ignored at this point. This allows the confounding between A, B, D, AB, AD, and BD to be resolved with eight experimental trials. The CE interaction is confounded with factor A, and the interaction between factors B and D. If neither of these is statistically significant, then the CE interaction will be explored. The full factorial experimental design for factors A, B, and D is given in Table 10.7.

Four of the experimental trials listed in Table 10.7 were run in the initial screening design, and each of these trials was run more than once. Thus, to complete the full factorial only requires four additional experimental trials. These four trials should be selected from the five-factor full factorial given in Table 10.2. The status of each of the experimental trials from Table 10.7 is given in Table 10.8.

Rows two and five from the $1/4$ fractional factorial were replicated to provide degrees of freedom for experimental error. Thus, there are 10 experimental trials from the full factorial which have been completed. These trials with the responses are shown in Table 10.9 along with the remaining four trials from the three-factor full factorial which have not been run.

The additional trials required to complete the full factorial for factors A, B, and D along with the trials from the $1/4$ fractional factorial are given in Table 10.10. Since

Table 10.8   Three-factor full factorial status.

Factor			Number of Times Run in $1/4$ Factorial
A	B	D	
−1	−1	−1	0
−1	−1	1	3
−1	1	−1	2
−1	1	1	0
1	−1	−1	3
1	−1	1	0
1	1	−1	0
1	1	1	2

Table 10.9   Three-factor full factorial results after completion of $^1/_4$ fractional factorial.

	Factor		
A	B	D	Noise
−1	−1	−1	
−1	−1	1	66.5
−1	−1	1	65.8
−1	−1	1	66.4
−1	1	−1	51.9
−1	1	−1	52.2
−1	1	1	
1	−1	−1	71.6
1	−1	−1	70.5
1	−1	−1	71.8
1	−1	1	
1	1	−1	
1	1	1	50.4
1	1	1	51.8

Table 10.10   Three-factor full factorial final results.

	Factor		
A	B	D	Noise
−1	−1	−1	66.0
−1	−1	1	66.5
−1	−1	1	65.8
−1	−1	1	66.4
−1	1	−1	51.9
−1	1	−1	52.2
−1	1	1	52.0
1	−1	−1	71.6
1	−1	−1	70.5
1	−1	−1	71.8
1	−1	1	70.6
1	1	−1	50.8
1	1	1	50.4
1	1	1	51.8

Table 10.11   Statistical analysis of the three-factor full factorial.

| Factor | Coefficient | Standard Error | p Value | 95% Confidence Interval | |
				Lower	Upper
Intercept	60.01	0.184	0.0%	59.56	60.46
A	0.94	0.184	0.2%	0.49	1.39
B	−8.52	0.184	0.0%	−8.97	−8.07
D	−0.03	0.184	88.8%	−0.48	0.42
AB	−1.48	0.184	0.0%	−1.93	−1.03
AD	−0.07	0.184	70.5%	−0.52	0.38
BD	0.09	0.184	64.3%	−0.36	0.54
ABD	0.16	0.184	41.7%	−0.29	0.61

the trials included in the $1/4$ factorial were run more than once, there are six degrees of freedom for experimental error. This further improves efficiency by eliminating the need for replications at this stage of the experiment, and improved accuracy.

Statistical analysis of the data in Table 10.10 yields factors A, B, and the AB interaction as statistically significant. This is shown in Table 10.11.

The confounding from the $1/4$ fractional factorial is shown in Figure 10.1 with the results of the statistical analysis from Table 10.11. A circle indicates statistical significance, and a slash indicates no statistical significance.

Note that each of the lines in the confounding decomposition diagram contains exactly one component that is statistically significant. This is the desired result. Before executing the three-factor full factorial, it was unknown if factor A was statistically significant or if one of the following components confounded with factor A was significant:

- interaction between factors B and D

- interaction between factors C and E

- interaction between factors A, B, C, D, and E

The results of the three-factor full factorial show factor A to be statistically significant and also show the BD interaction not to be statistically significant. If

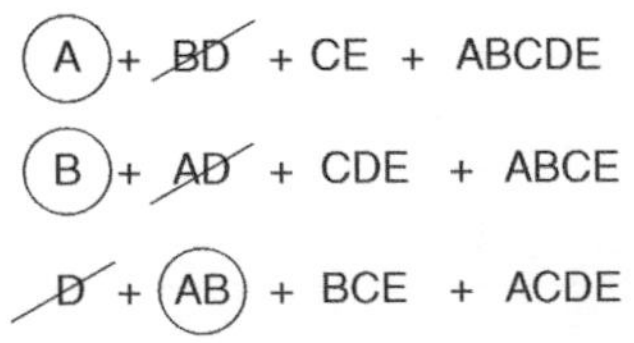

Figure 10.1   Confounding decomposition diagram.

neither factor A nor the BD interaction were statistically significant, then more trials from the five-factor full factorial would have been required to determine whether the CE interaction or the ABCDE interaction was statistically significant.

The same is true for the remaining two lines in Figure 10.1. The second line shows factor B, in the $^1/_4$ fractional factorial, is responsible for the change in the response, and that the AD interaction did not have a statistically significant impact on the response. The third line shows factor D did not impact the response, and that the interaction between factors A and B has a statistically significant impact on the response.

It is possible that more than one of the confounded components is statistically significant. For example, factor A and the interaction between factors C and E could both be statistically significant. A review of the estimated coefficients provides insight to this issue. The confidence intervals for the estimated coefficient of A in the fractional factorial (Table 10.6) and the full factorial (Table 10.11) overlap indicating no statistically significant difference. The coefficient estimated in the fractional factorial is the sum of the coefficients for factor A, the BD interaction, the CE interaction, and the ABCDE interaction. Since the coefficients from both experiments are statistically equivalent, the effects of the BE interaction, the CE interaction, and the ABCDE interaction are either close to zero or equal in magnitude and opposite in direction, which is unlikely. The coefficients for the terms B and AB are also statistically equal, confirming the confounding terms have no impact or a very small effect.

In summary, 14 experimental trials were required to determine factors A, B, and the interaction between them were statistically significant. This was done in a sequential fashion by purposely beginning with as much confounding as possible, and then separating the confounded components. Using a full factorial design would have required 32 trials and at least two replications for experimental error. Using a sequential approach resulted in a 59% reduction in the number of experimental trials.

10.2   Response surface

A response surface experiment extends the screening design to allow estimation of quadratic effects. The form of the two-level empirical model after resolving confounding is

$$
\begin{aligned}
y = c_0 &+ c_1 X_1 + c_2 X_2 + \cdots + c_n X_n \\
&+ c_{12} X_1 X_2 + c_{13} X_1 X_3 + \cdots + c_{(n-1)(n)} X_{n-1} X_n
\end{aligned}
\tag{10.1}
$$

The response surface experiment expands this equation by adding quadratic or even cubic terms. The form of the empirical model with quadratic terms is

$$
\begin{aligned}
y = c_0 &+ c_1 X_1 + c_2 X_2 + \cdots + c_n X_n \\
&+ c_{12} X_1 X_2 + c_{13} X_1 X_3 + \cdots + c_{(n-1)(n)} X_{n-1} X_n \\
&+ c_{11} X_1^2 + c_{22} X_2^2 + \cdots + c_{nn} X_n^2
\end{aligned}
\tag{10.2}
$$

Table 10.12   Trials required to estimate coefficients for a quadratic model.

Number of Factors	3^n Full Factorial Number of Trials	Quadratic Model Number of Coefficients
2	9	6
3	27	10
4	81	15
5	243	21
6	729	28
7	2187	36

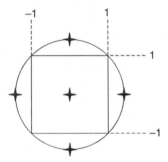

Figure 10.2   Circumscribed central composite (CCC) response surface design.

When estimating the coefficients of the model displayed in Equation 10.2, fractional factorials with three levels are recommended. A full factorial with three levels includes enough trials to estimate all interactions, but quickly becomes large as the number of factors increases. Table 10.12 compares the number of trials required for a three-level full factorial experiment to the number of coefficients that must be estimated for a quadratic model.

10.2.1   Central composite designs[4]

A Box-Wilson central composite design, commonly called a central composite design, contains an imbedded factorial or fractional factorial matrix with center points augmented with a group of star points that allow estimation of curvature. A central composite design always contains twice as many star points as there are factors in the design. There are three general types of central composite designs.

Circumscribed, or CCC, designs are the original form of the central composite design. The star points are located on a circle encompassing a screening design. This is shown in Figure 10.2 using −1 and 1 as the low and high levels of the original

---

[4]The accompanying website contains CCF designs for up to 10 factors. The file name is CCF.xls.

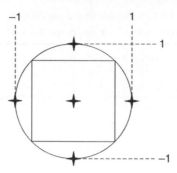

Figure 10.3   Inscribed central composite (CCI) response surface design.

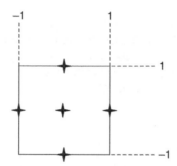

Figure 10.4   Face-centered central composite (CCF) response surface design.

screening design with two factors. It may not be possible to augment a screening design, represented by the corners of the square, if the levels of the screening design are absolute minimum and maximum, as the star points exceed these values.

For an inscribed central composite design, CCI, the star points are set at the minimum and maximum levels. If augmenting a screening design, the levels of the screening design will be within the absolute limits to allow the expanded settings for the star points. This is shown in Figure 10.3.

For a face-centered central composite, CCF, the star points are located on the square (for two-factor designs) formed by the screening design. This design is not rotatable, and has slightly less precision for coefficient estimation. Its advantage is that star points can easily be added to prior screening designs (Figure 10.4).

Example 10.2[5]

Augment the screening design presented earlier in this chapter with additional points to allow a second-order polynomial to be created.

---

[5] This example is contained on the accompanying website. The file name is AutomotiveNoise.xls.

Table 10.13   Additional experimental trials (coded) required for the response surface.

Factor		
A	B	Noise
0	−1	77.6
0	0	68.8
0	1	78.8
1	0	52.6
0	−1	80.7

Solution

Factors A and B are significant. A CCF design with two factors requires nine trials. Table 10.10 displays the last results from the screening design. Four of the trials required for a CCF have been completed with 10 replications. The five trials in coded units that must be completed for the CCF design with output results are given in Table 10.13.

The data from Tables 10.10 and 10.13 can be used to estimate the coefficients of a second-order polynomial. Before the regression can be executed, columns must be included for each term in the polynomial. Un-coded data will be used for the statistical analysis to allow development of an equation that uses un-coded data as an input. The terms A^2, B^2, and AB are computed from the values of A and B. This is shown in Table 10.14.

Table 10.15 shows the statistical analysis for the data in Table 10.14. The A^2 term is not statistically significant (p value greater than 5%), which indicates there is no curvature in the model in relation to factor A.[6] Terms that do not contribute to the model should not be included in the model, so the regression will be repeated excluding A^2.

The regression results after removing the nonsignificant terms are given in Table 10.16.

The resulting empirical model is

$$Y = -186.7 + 2.006A + 93153B - 8384127B^2 - 305.6AB$$

The precision of the empirical model can be accessed by constructing a residual plot. The residuals, shown in Figure 10.5, are the difference between the measured noise and the noise predicted by the empirical model. The residual plot indicates the

---

[6] Caution should be used when making decisions from a statistical analysis conducted using un-coded data. The differences in the order of magnitude of the factors may distort the results. In this case, the results are confirmed by repeating the regression analysis using coded units.

Table 10.14   Response surface data.

| Factor or Term | | | | | |
A	B	A^2	B^2	AB	Noise
18.33	0.0040	336.0	0.00001600	0.0733	66.0
18.33	0.0040	336.0	0.00001600	0.0733	66.5
18.33	0.0040	336.0	0.00001600	0.0733	65.8
18.33	0.0040	336.0	0.00001600	0.0733	66.4
18.33	0.0070	336.0	0.00004900	0.1283	51.9
18.33	0.0070	336.0	0.00004900	0.1283	52.2
18.33	0.0070	336.0	0.00004900	0.1283	52.0
24.81	0.0040	615.5	0.00001600	0.0992	71.6
24.81	0.0040	615.5	0.00001600	0.0992	70.5
24.81	0.0040	615.5	0.00001600	0.0992	71.8
24.81	0.0040	615.5	0.00001600	0.0992	70.6
24.81	0.0070	615.5	0.00004900	0.1737	50.8
24.81	0.0070	615.5	0.00004900	0.1737	50.4
24.81	0.0070	615.5	0.00004900	0.1737	51.8
18.33	0.0055	336.0	0.00003025	0.1008	77.6
21.57	0.0040	465.3	0.00001600	0.0863	68.8
21.57	0.0055	465.3	0.00003025	0.1186	78.8
21.57	0.0070	465.3	0.00004900	0.1510	52.6
24.81	0.0055	615.5	0.00003025	0.1365	80.7

empirical model is accurate to ± 1 decibel. Be aware that the precision of the model may be worse at factor levels not tested.

If the empirical model is used to estimate output variance using the Taylor series approximations described in earlier chapters, the precision of the variance estimate can only be accessed through experimentation. If the precision of the estimated

Table 10.15   Response surface regression results (un-coded data).

Factor	Coefficient	Standard Error	p Value	95% Confidence Interval	
				Lower	Upper
Intercept	−200.7	15.96	0.0%	−235.2	−166.2
A	3.417	1.505	4.1%	0.165	6.669
B	92 803	1902	0.0%	88 694	96 912
A^2	−0.03271	0.0347	36.3%	−0.1076	0.0422
B^2	−8 352 632	161 734	0.0%	−8 702 036	−8 003 227
AB	−305.6	31.37	0.0%	−373.4	−237.9

Table 10.16   Response surface regression results after removing A^2.

Factor	Coefficient	Standard Error	p Value	95% Confidence Interval	
				Lower	Upper
Intercept	−186.7	5.807	0.0%	−199.1	−174.2
A	2.006	0.1716	0.0%	1.6377	2.3737
B	93 153	1858	0.0%	89 167	97 139
B^2	−8 384 127	157 634	0.0%	−8 722 218	−8 046 036
AB	−305.6	31.24	0.0%	−372.7	−238.6

nominal output, or the estimated variance is not sufficient, there are two options for improvement:

- use a higher-order polynomial or

- reduce the size of the region modeled.

A third-order polynomial can be obtained by adding four additional points as shown in Figure 10.6. The solid points in Figure 10.6 are from previous experiments, so only four additional experiments are required.

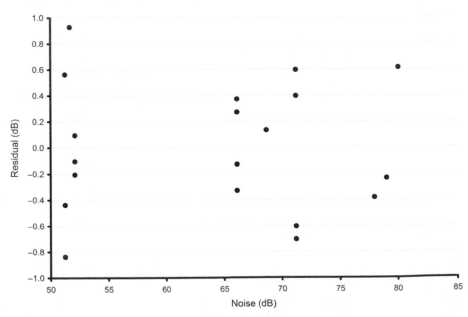

Figure 10.5   Empirical model residual plot.

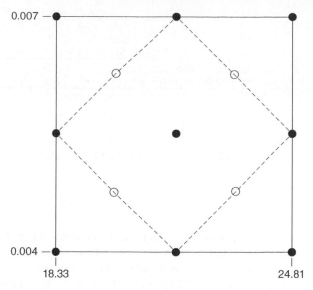

Figure 10.6   Third-order polynomial experimental design.

By adding the five open points in Figure 10.7, a second-order polynomial can be estimated for a region one-fourth the size of the original region. This procedure can be repeated, allowing second-order polynomials to be estimated for smaller and smaller regions until the desired accuracy is reached.

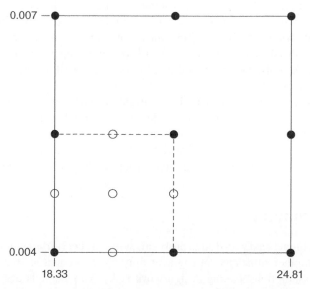

Figure 10.7   Reduced region for second-order polynomial.

				Outer array			
			Salt	−1	−1	1	1
			Runout	−1	1	−1	1
			Temperature	1	−1	−1	1
	Inner array						
Trial	Viscosity	Roughness	% Manganese	R_1	R_2	R_3	R_4
1	−1	−1	1				
2	−1	1	−1				
3	1	−1	−1				
4	1	1	1				

Figure 10.8   Inner and outer array example.

10.3   Taguchi

Running each trial of an experiment allows an estimate of the mean. If each trial is run more than once, the standard deviation can also be estimated. The result is two empirical models: one to estimate the mean and one to estimate the standard deviation. This technique is often referred to as "Taguchi methods" as Genichi Taguchi has been advocating this approach since the 1950s.

When replicating experimental trials to obtain an estimate of the standard deviation, it is important to include all sources of variation. If trials are replicated without regard for sources of variation, the standard deviation will be underestimated. A P-diagram (introduced in Chapter 3) can be used to identify sources of variation and ensure they are included in the experiment.

Taguchi refers to the inner and outer array. The inner array is the same as the experimental designs discussed earlier. The outer array is used to ensure noise factors are properly introduced into the experiment. Figure 10.8 shows an example of a three-factor design with an outer array containing three noise factors. In this design, each of the four trials is replicated four times. For example, the first replication (R_1) is conducted with salt and runout at the lower level, and temperature at the high level. The levels for the noise factors should be close to the lowest and highest values the design will experience.

The number of replications used is critical to accuracy. In general, there is more error when estimating the standard deviation than when estimating the mean. Figure 10.9 shows 95% confidence intervals for standard deviation as a function of sample size, given a sample standard deviation of 1.0. As seen in Figure 10.9, at least four replications are required to bring the estimate error to less than 300%.

10.4   Summary

Engineers are often faced with design problems without access to equations describing the system. In other situations, an engineer may have an equation, but the equation is based on assumptions making design work risky. The methods described in this chapter can be used to efficiently develop new empirical models, or empirical models

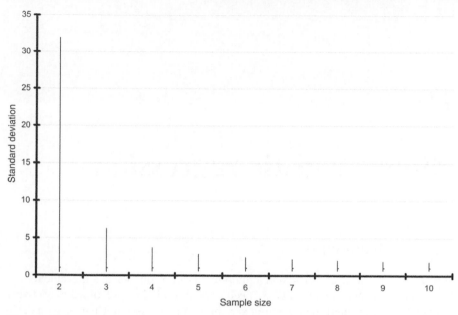

Figure 10.9 Confidence intervals for standard deviation as a function of sample size.

to enhance existing models. The models can then be used for robust design as described in earlier chapters.

Chapter 8 describes how numerical methods can be used to obtain a robust design when black boxes are used for engineering modeling. An alternative to this is to use the techniques presented in this chapter to build an empirical model from the black box. Statistical significance cannot be used with this approach since there is no experimental error (all replication results give the same result). Significance must be determined using the estimated coefficient or experimental effect.

Exercises

All exercises are contained in the DOE trainer software. This software provides simulations of engineering systems, and is located on the accompanying website.

11

Binary logistic regression

11.1 Introduction

In the conversion of end-customer requirements into functional requirements and ultimately to design requirements, we need to establish linking relationships between explanatory input and response output variables. As discussed in Chapter 6 on desirability, these response outputs may include numerical measures such as customer satisfaction, usability, performance, and cost or categorical outputs such as *will purchase* versus *will not purchase* or *has* versus *does not have* features (binary outcomes). To meet desirability goals for response outputs, we must identify critical explanatory input variables along with settings or a range of settings.

A recurring challenge in establishing useful linking relationships is that they may change, possibly even from a linear to a quadratic relationship, depending on the range of the explanatory variable. To illustrate, consider the square footage of homes. Suppose several houses in a particular area with similar characteristics are selected, and the square footage versus selling price is compared. Here, you might find the following relationship as shown in Figure 11.1.

In the first region of the figure, the relationship between square footage and price appears fairly flat. Then, we have a range in which there is a fairly steep linear relationship. Finally, the relationship changes where incremental increases in square footage begin to level off and thus there is only a marginal relationship between square footage and the selling price. This relationship may be referred to as an S-curve (Hilbe, 2009).

Several implications may be drawn from this simple example. The relationship between explanatory and response variables often changes depending on the range of variation for the explanatory variable studied. In some regions, this relationship may be significant (steep slope), whereas for other ranges, the relationship may be

Probabilistic Design for Optimization and Robustness for Engineers, First Edition.
Bryan Dodson, Patrick C. Hammett and René Klerx.
© 2014 John Wiley & Sons, Ltd. Published 2014 by John Wiley & Sons, Ltd.
Companion website: http://www.wiley.com/go/robustness_for_engineers

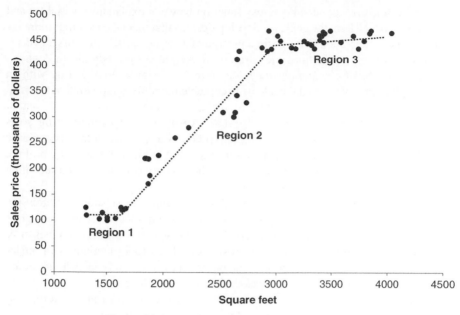

Figure 11.1   Home price versus square footage.

relatively flat. In product design, a flat region may be referred to as a *robust operating range* as the response is less sensitive to explanatory input variable within the range of study.

These regions have clear implications in terms of product cost. For instance, one cannot continuously reduce the square footage to reduce the selling price, nor should one expect that by simply increasing the square footage, customers are willing to pay more for a home at the same price per square foot.

We may extend this concept to customer loss functions. A *customer loss function* represents the relationship between an explanatory design input variable setting, or range of settings, and a customer measure of desirability such as customer satisfaction, usability, performance, and cost. For example, with a customer satisfaction metric, designers assume that target customers have an optimal value (meeting a desired usability and performance for a particular price) in which a large percentage of customers will be satisfied with the product's value. Similarly, they often assume a value or region exists in which if the explanatory variable is not achieved (e.g., below a lower limit in the larger-the-better case), the customer will become dissatisfied and have a desirability of zero (product with no value).

To complicate the establishment of these relationships, some trade-offs are inevitable between response output metrics. These trade-offs may occur among dependent performance metrics such as engine power and fuel economy. In fact, achieving higher performance levels often entails using higher cost raw materials or more advanced system architecture, technology, or processing equipment. As such, design decisions related to material selection, technology, and system architecture are greatly aided by establishing a loss function between explanatory variables and response outputs. Here, the goal is to develop designs that meet the user requirements and customer expectations for a desired portion of a target market. Of course, in dealing with customer satisfaction, one is rarely, if ever, able to satisfy all. In contrast, *"trying to satisfy all can end up satisfying none"* as designers may end up with an over-designed product priced outside its acceptable range for a target market.

Modeling the relationship between explanatory inputs and response output variables is a fundamental engineering skill. An often used method to do so is ordinary least squares regression. It is used to predict values of a continuous response variable using one or more explanatory variables, and to identify the strength of the relationships between these variables. Thus, the two goals of regression are often referred to as prediction and explanation.

Numerous techniques have been developed to apply least squares regression (simple, multiple, best subsets, stepwise, weighted, polynomial, etc.) and work quite well if the response output is an interval or ratio-scaled variable (Hutcheson & Sofroniou, 1999). Still, there are many customer related loss functions with binary categorical responses. For instance, customers may ultimately decide to either purchase or not purchase a product based on key product features.

Relatedly, companies routinely survey customers on ease-of-use and customer satisfaction. While these surveys often employ Likert-type questions (Smith & Albaum, 2009) with interval response ratings from say one to five, they are essentially ordinal-scaled categories. Thus, they have an implied order but not a clear magnitude of difference property between rating values. In addition, given the relatively few response levels for such survey questions, one rarely use them to draw clear relationships to establish a customer loss function. As such, companies routinely collapse the responses into acceptable and unacceptable (or satisfied versus not satisfied). For instance, a five-point scale may be collapsed into a binary response by assigning ratings from one to three a value of zero, and four and five ratings a value of one.

For a binary response, a more appropriate and powerful tool for establishing linking relationships is binary logistic regression. In this chapter, we provide a brief overview of this technique including a comparison with ordinary least squares regression. Then, we demonstrate the application of binary logistic regression with different types of explanatory input variables including binary inputs and numerical input variables using a study on automotive fuel filler door complaints for a new vehicle design application. With this example, we discuss testing for a statistically significant relationship for explanatory input variables and the establishment of an odds ratio to measure the strength of association between the inputs and binary output. Next, we show how to establish prediction models and event probability plots to create a customer loss function between those explanatory inputs with the strongest association. Finally, we repeat the approach for a special case where the response is not a binary output but a maximum (or minimum) value response point in which a customer indicates dissatisfaction. For this section, we use an example from a steering column adjustment study to determine the maximum amount of reach before target customers are predicted to become dissatisfied.

11.2   Binary logistic regression

Let us begin with a review of binary logistic regression. In this section, we discuss different types of logistic regression, why binary logistic regression often is preferred over ordinary least squares regression for binary outputs, and why the logit function may be used to transform data to relate an explanatory variable to the probability of a successful binary event.

11.2.1   Types of logistic regression

Regression analysis does not require a continuous, numerical output. In fact, numerous examples exist where the output for a study is categorical where we may apply logistic regression (Hilbe, 2009; Menard, 2010). Some categorical response outputs include:

- pass or fail,

- make or buy,

- purchase or do not purchase,

- satisfied or not satisfied,

- amount of lubrication applied (low, medium, high), and

- medication delivery (wrong medication, wrong administer time, wrong amount).

In the first four of these output examples, the response is binary where it is customary to code responses as zero or one. For these, we may apply binary logistic regression. In the example with lubrication amount, we have an ordinal rating scale. Here, we may apply ordinal logistic regression. However, as mentioned above with the Likert survey example, we may also compress the scale to a binary response such as *low* versus *medium* or *high*. We often compress ordinal scales with relatively few levels into binary scales to simplify the analysis, especially if the ultimate objective is to determine acceptable versus unacceptable performance.

For the medical delivery process example, suppose we may classify defects into more than two unordered nominal categories (e.g., wrong medication, wrong time for medication delivery, and wrong amount of medication). Here, we have a nominal output and may apply nominal logistic regression. A limitation in applying nominal logistic regression, however, is that it typically requires sample sizes larger than are practical for product development decisions (Agresti, 1996). The majority of logistic regression analysis is done using a binary output, thus, we focus this chapter on this type of analysis.

To use binary logistic regression, we first define a *successful event* and code their occurrences as one. For example, we may define a successful event as an acceptable part. A *successful* event is not necessarily a positive outcome. In fact, we often define success in logistic regression using a negative outcome such as a defective part or a customer complaint. Here, we may be interested in finding the causes of defects or sources of customer complaints. In this latter situation, we may assign all customers that complain a value of one, and all those that do not complain a zero. We also should note that a binary outcome may not necessarily be positive or negative. For instance, suppose we wish to predict gender based on anthropometric characteristics. Here, we may arbitrarily assign *male* a value of one and *female* a value of zero.

Even if response data are interval or ratio scaled, we still may assign a cutoff value to define a binary outcome. This is commonly done in manufacturing using specification limits. If a part dimension is within its specification limits, we may code responses as zero, and assign a code of one for anything outside the limits. Similarly, if using a seven-point ordinal-interval survey rating scale, we may assign a five or higher as acceptable and anything else as unacceptable. In some cases, identifying the acceptance limits or cutoff values may be straightforward, other times we must rely on experience or expert opinion.

Once we code responses as either a zero or one, we then may estimate the mean of the distribution. For binary response, this is simply the proportion of ones in the sample. For instance, if 100 customers are surveyed and 20 complain about a product's lack of usability, the mean or proportion of complaints (coded value = 1) in the distribution would be 0.2. Thus, if we randomly select a customer from this sample, the probability that this customer will complain is 0.2. Here, the proportion, p, is equivalent to the probability. Of note, the proportion of zeros is $1 - p$ and is sometimes denoted using the letter q where $q = 1 - p$.

11.2.2   Binary versus ordinary least squares regression

To illustrate why we often prefer binary logistic over ordinary least squares regression, we may use a classic example where one wishes to predict gender given a person's height. Suppose we sample 128 people, and record their height and gender (see Figure 11.2). We then create a scatter plot with a regression trend line in the form

$$y = \beta_0 + \beta_1 x \tag{11.1}$$

With this simple example, we may identify several concerns with applying least squares regression. First, the binary response only may take on two values (0 and 1). Thus, the model offers little value other than predicting the average standing height for a female (156.5) and the average standing height of a male (180.8). In addition, the predicted values for the response may be greater than one or less than zero, which is not physically possible.

In similar binary cases, we may be violating several assumptions of least squares regression. For instance, the residual errors are not normally distributed, and the variance of the dependent variable is not constant across the values of the independent variables. To demonstrate, the variance of a binary response may be estimated as the product of p and q. If the proportion equals 0.5, the variance is maximized at 0.25, and as p approaches zero or one, the variance approaches zero and thus is not constant over the range of values for the independent variable (Pohlman & Leitner, 2003).

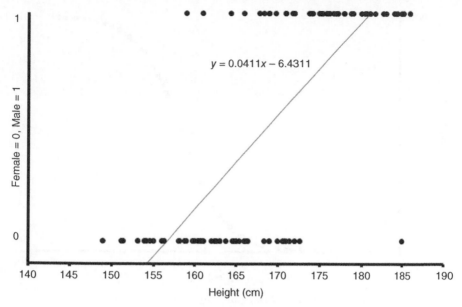

Figure 11.2   Gender versus height.

To better depict the relationship between height and gender, suppose we compare height to the proportion of males (Figure 11.3). A proportion of zero would indicate all females and a proportion of one would indicate all males.

Several observations may be made from this graph. A best fit line represents a rolling average, just as in linear regression. In addition, the relationship is not linear, flattening at small and large values of height.

Here, rather than a linear relationship, we may fit a logistic curve to relate the independent variable to the proportion of ones. The proportion of ones used as the dependent variable is

$$p_i = \frac{e^{(\beta_0 + \beta_1 x_i)}}{1 + e^{(\beta_0 + \beta_1 x_i)}} = \frac{1}{1 + e^{-(\beta_0 + \beta_1 x_i)}} \tag{11.2}$$

where p_i is the proportion of ones for the ith point, β_0 is the intercept resulting from linear regression of the binary output versus the input, and β_1 is the intercept resulting from linear regression of the binary output versus the input.

The constant, β_0, yields a probability of zero when independent variable is zero, and the coefficient β_1 represents how quickly the probability changes when the independent variable is changed by a single unit. A positive value for β_1, as shown with these standing height data, suggests an increasing function while a negative value for β_1 would indicate a decreasing function.

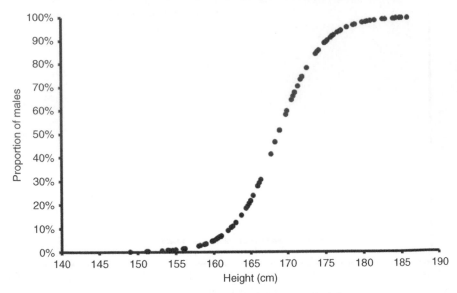

Figure 11.3   Proportion male versus height.

11.2.3   Binary logistic regression and the logit model

Another useful observation may be made from the binary logistic curve. A hidden linear model may be revealed through a proper data transformation. A classic data transformation for such a situation is the logit function. This function may be expressed as follows:

$$\ln\left(\frac{p_i}{1 - p_i}\right) = \beta_0 + \beta_i x_i \tag{11.3}$$

Using binary logistic regression model parameters, β_0 and β_1, and the specific values for height, we may create a scatter plot (Figure 11.4) using the logit function as the dependent variable. A linear relation results from the transformed data.

Thus, the logit function has been adjusted for the curved relationship between the independent variable and the proportional response as well as restricted all predicted values for p between zero and one. Still, we have not resolved issues with constant variance and normality assumptions for the residuals. As such, we cannot use traditional measures to assess goodness of the model fit nor statistical significance. Furthermore, there is no closed mathematical solution to estimate the regression model parameters.

Instead, binary logistic regression relies on iterative numerical analysis methods to generate maximum likelihood estimates for the model parameters and related methods to assess statistical significance of the explanatory input variables. We leave these derivations to other references (Dixon et al., 2000). For this book, we assume

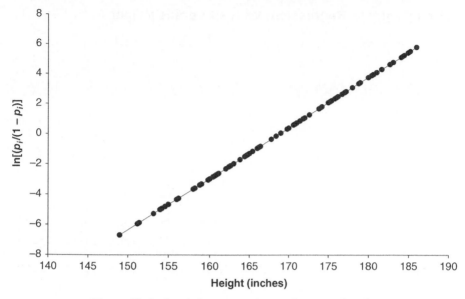

Figure 11.4   Logit for proportion male versus height.

that users will utilize an advanced statistical software package to generate the model parameters and provide statistical output to assess statistical significance.

For instance, the results of binary logistic regression between height and gender using a commercial software package are shown in Figure 11.5.

This software provides estimates of the model coefficients along with the ability to assess statistical significance. From here, the conventional approach is to conclude a statistically significant effect for an independent variable if the p value from the software is less than 0.05 (0.05 equates to 95% confidence). Here, the p value for standing height is approximately zero, so we conclude, as expected, a significant relationship. The statistically significant p value for the constant term implies the constant differs statistically from zero.

With a logistic regression model, the coefficients often are difficult to interpret because the effect of increasing or decreasing the independent variable, by one unit, varies based on the value of the independent variable. Rather than communicate the effect of an input variable using its coefficient, we may convey results using odds or an odds ratio. For instance, if the probability of an event is 0.75, the odds are three to one, or three times more likely to occur.

With logistic regression, we typically model the log-odds, which may be expressed as

$$e^{p_i} = \frac{p_i}{1 - p_i} \tag{11.4}$$

Binary Logistic Regression: Male = 1 versus Height

```
Link Function: Logit


Response Information

Variable  Value  Count
Male = 1  1         63   (Event)
          0         65
          Total    128


Logistic Regression Table

                                             Odds      95% CI
Predictor       Coef     SE Coef      Z      P  Ratio  Lower  Upper
Constant     -56.9419    9.63908   -5.91  0.000
Height       0.337292  0.0570413    5.91  0.000  1.40   1.25   1.57


Log-Likelihood = -41.032
Test that all slopes are zero: G = 95.351, DF = 1, P-Value = 0.000


Goodness-of-Fit Tests

Method             Chi-Square  DF     P
Pearson              271.504   75  0.000
Deviance              49.105   75  0.991
Hosmer-Lemeshow       35.487    8  0.000
```

Figure 11.5   Binary logistic regression results using statistical software.[1]

Using the example above, the odds ratio for height is 1.4 ($e^{0.337}$). The positive coefficient yields an odds ratio greater than one, suggesting an increasing relationship between height and the probability of being a male. In contrast, if the coefficient is negative, the odds ratio will be less than one and thus we would have a lower likelihood of the event occurring with a unit change in the independent variable.

An interesting result for an odds ratio occurs if the coefficient is not shown to differ significantly from zero. For a coefficient of zero, the odds ratio is one (e^0). This is sometimes viewed as the baseline odds. Here, we have a presumption of no marginal effect with a change in the independent variable. Stated another way; we have an equal likelihood observing the event for any value of the observed range.

---

[1]Data for this example are contained on the accompanying website. The file name is "Gender_Height _Data.xlsx."

From another perspective, if the 95% confidence interval shown in the above statistical software output for the odds ratio includes one, we may infer that the independent variable has no marginal effect.

11.2.4   Binary logistic regression with multiple predictors

As with any regression analysis, we may wish to create a binary logistic model to predict a response given multiple explanatory or predictor variables. The general multiple binary logistic regression model is given in Equation 11.5.

$$p_i = \frac{1}{1 + e^{-(\beta_0 + \beta_1 x_{1i} + \beta_2 x_{2i})}} = \frac{1}{1 + e^{-(-6.233 + (0.058)(63) + (1.789)(1))}} = 0.3122 \quad (11.5)$$

As with multiple regression analysis, we seek to develop the most simplistic binary logistic model (fewest input variables) that explains the most variation in the dependent variable. As such, we first may utilize more classical techniques, such as best subsets regression or stepwise regression to filter out noncritical independent variables. While we may not be able to succinctly define a single useful model, we should consider the primary objective, which is to develop a customer loss function for designers. As such, we need to combine statistical as well as nonstatistical reasoning and practical limitations.

As another consideration, if the input variables are not independent of each other, we have a condition known as multicollinearity (Doane & Seward, 2007). As with multiple regression, we need to be mindful of potential problems with multicollinearity (Montgomery et al., 2012).

11.2.5   Binary logistic regression and sample size planning

One reason for seeking simple, yet practically useful, binary logistic regression models is related to power and sample size planning. The use of proportions generally requires a much larger sample size compared with interval or ratio-scaled data. In addition, logistic regression relies on numerical analysis methods and maximum likelihood estimation, which require a sufficient sample size to generate the coefficient estimates. Thus, sample size planning is particularly important to binary logistic regression analysis.

With binary logistic regression, sample sizes above 500 are usually adequate while sample sizes less than 100 may potentially be risky unless only large effects are of interest. Collecting data is costly and time-consuming, particular during the design phase of new product development, which puts pressure to reduce sample size. Alternatively, to obtain useful models with fewer samples, Long (1997) recommends the following guidelines for identifying sample size requirements.

$$N = 50 + (10)(\text{number of model predictors}) \quad (11.6)$$

With three predictors, the above equation suggests a minimum sample size of 80. These guidelines aside, the sample size for a binary logistic regression study ultimately may be driven by what is available rather than what is desired. For instance,

Table 11.1   Binary classification of fuel fill door performance.

PPH Group	Description	Number of Vehicles	Percentage of Vehicles
1	Two consecutive years with PPH > 0.5	16	19.5%
0	Others	66	80.5%

if you are conducting an analysis to examine the relationship between an automotive vehicle design characteristic and customer satisfaction, the available sample size may be limited to the number of vehicles in a particular market segment. Here, we are limited to the best available data.

11.2.6   Binary logistic regression fuel door example

To explore the use of logistic regression, we consider a project to identify potential factors resulting in low JD Power[SM] performance related to fuel filler doors. Here, survey complaint data were obtained from 82 available vehicles.[2]

The first step in any analysis is problem definition. In this study, the objective is to identify design characteristics on existing vehicles that are potentially associated with high customer complaints so future designs minimize the likelihood of complaints. As an aside, we certainly recognize that customer complaint data are only one factor to be considered in such design decisions and thus the results shown here represent only one consideration of the final design recommendations.

The external survey data used in this study are from an average of 400 customers per year, per each vehicle. Customers answered a series of questions on fuel filler doors, with the results converted into the metric, problems per hundred vehicles (PPH). Unfortunately, this survey information has relatively low discrimination to run an ordinary least squares regression analysis using PPH as the response, thus, the team decided to categorize vehicles into a binary classification based on the distribution of historical PPH compliant data and analyst experience. The team established the defect group to include vehicles having two consecutive years with a PPH greater than 0.5. Using this classification, we may classify the 82 vehicles per Table 11.1. The response output variable is denoted as *HighPPH-1* for this example.

As discussed previously, an issue with coding binary responses is identifying which outcome to assign a value of one. While technically it is irrelevant, practically it is recommended to assign a one based on the outcome of interest. If a study is intended to identify things gone wrong, we recommend assigning a one to the negative outcome or defect. In contrast, if the objective is to study things gone right in a design, we recommend coding the positive outcome a value of one. Again, the primary purpose for this guideline is related to the presentation of results to others. Here, we assign a value of one to the negative outcome of interest.

---

[2] The data are contained on the accompanying website. The file name is "JDPower_Data.xlsx."

Table 11.2   Explanatory input variables for fuel fill door study.

Input Variable	Type	Description
Locking door	Binary	Fuel door locks (No = 0, Yes = 1)
Feature line	Binary	Body has feature line through fuel filler door that is offset from center line of door (= 1)
One click	Binary	Close cap with one click (code = 1) or ratchet (code = 0)
Pocket width	Continuous	Width of the fuel pocket opening
Cap depth	Continuous	Depth of the fuel pocket to the fuel cap
Body angle	Continuous	Angular measurement of the cap for the fuel relative to the fuel filler door

Next, we need to identify potential explanatory variables. In this study, numerous vehicle characteristics were studied. We will focus on the following subset of explanatory or design input variables in Table 11.2.

We may use these input variables to illustrate different analyses to demonstrate the usage of binary logistic regression and the development of a customer loss function. These include:

- compare a statically significant binary input with a binary response,

- compare a nonsignificant binary input with a binary response,

- compare a continuous input variable with a binary response, and

- develop a final model with multiple explanatory input variables.

11.2.7   Binary logistic regression—significant binary input

We begin the analysis of the fuel filler door by examining the relationship between the binary input, *feature line*, and the response output variable, *HighPPH-1*. Feature line is a binary classification developed to distinguish between vehicles that have a feature line through the filler door on the vehicle body offset from the filler door center line (offset may be high or low) versus those vehicles either having no feature line or a feature line through the center of the vehicle. We appreciate that we have compressed levels for the independent variables. Given the limited available sample size, the team reasoned that this categorization was reflective of the potential concern of interest, an asymmetrically appearing feature line through the door.

To assess the statistical significance of this *feature line* factor, we may use the binary logistic regression function within any advanced statistical software package. To run a binary logistic regression analysis, we need to identify the response (*HighPPH-1*) and a model variable (*feature line*). We also recommend using the logit function and request additional analysis for goodness-of-fit tests and measures of association. The binary logistic regression analysis using the logit function results are shown in Figure 11.6.

Binary Logistic Regression: HighPPH-1 versus Feature Line

```
Link Function: Logit


Response Information

Variable   Value  Count
HighPPH-1  1         16  (Event)
           0         66
           Total     82


Logistic Regression Table

                                                Odds      95% CI
Predictor           Coef    SE Coef      Z      P  Ratio  Lower  Upper
Constant         -1.77307  0.360526  -4.92  0.000
Feature Line      1.15403  0.591404   1.95  0.051   3.17   0.99  10.11
```

Figure 11.6   Binary logistic regression results for HighPPH-1 versus feature line.

From this analysis, we observe a p value of 0.051. Although the conventional criterion is to conclude a statistically significant effect if the p value is less than 0.05, we recommend using 0.10 for at least an initial analysis of individual factors with binary logistic regression. Since binary logistic regression uses iterative numerical analysis to assess significance, p value estimates are approximations.

We also observe a positive coefficient of 1.154 for this example with an odds ratio of 3.17 (odds ratio $= e^{1.154}$). Thus, the likelihood of being in the high defect group is approximately three times higher if the vehicle design includes a feature line through the fuel fill door that is offset from the center line.

11.2.8   Binary logistic regression—nonsignificant binary input

Next, we consider the case for the input variable *locking door*. We examine this effect using a binary logistic regression analysis, and the results are given in Figure 11.7. The variable *locking door* does not have a statistically significant effect on high PPH complaints (p value $= 0.828$). Furthermore, the 95% confidence interval for the odds ratio includes one suggesting this variable has no marginal impact.

11.2.9   Binary logistic regression—continuous input

Next, we examine the case of a numerical continuous input. Ideally, when running a binary logistic regression to create a loss function, we prefer to have one or two key continuous input variables. While this may not always occur or even be practical, we urge practitioners to try and identify such variables to give the designer a range of settings versus a single point solution or recommendation from a binary or nominal design characteristic.

Binary Logistic Regression: HighPPH-1 versus Locking Door

```
Link Function: Logit


Response Information

Variable   Value  Count
HighPPH-1  1         16   (Event)
           0         66
           Total     82


Logistic Regression Table

                                                Odds      95% CI
Predictor          Coef   SE Coef      Z      P Ratio  Lower  Upper
Constant       -1.47591  0.391882  -3.77  0.000
Locking Door    0.121361  0.557521   0.22  0.828  1.13   0.38   3.37
```

Figure 11.7   Binary logistic regression results for HighPPH-1 versus locking door.

To illustrate the analysis with a continuous input, we consider the variable, *body angle*. This angle measures the orientation of the fuel cap relative to the car body, with smaller angles typically desired. For numerical variables, we may visualize their relationship to the response with a scatter plot as shown in Figure 11.8.

Several observations may be gleaned from this figure. All of the observed vehicles with an angle less than 60° fell in the low PPH group. Several vehicles with larger angles also fell in the low PPH group as well. Clearly, we cannot simply recommend a body angle criterion that will ensure a vehicle will fall in the low PPH group. However, we may establish a likelihood of falling into each group associated with this measurement. As stated previously, numerous factors may affect complaints and we are not presuming true causal relationships here. Rather, we are looking to establish associated probabilities or potential loss based on the best information available.

If we run a binary logistic regression using body angle by itself (Figure 11.9), we may observe this variable is not statistically significant (*p* value = 0.451). However, given the above scatter plot and the additional fact that other variables such as feature line likely have a significant effect, we will not exclude this variable yet. As with classic regression, developing a model given multiple explanatory inputs, we need to consider the collection of variables as a system.

11.2.10   Binary logistic regression—multiple inputs

As with classic regression, to build a useful prediction model, we often need to consider multiple factors. Unfortunately, binary logistic regression tends to require larger sample sizes. As such, we recommend using classical methods to first narrow down the number of explanatory input factors prior to creating a final logistic

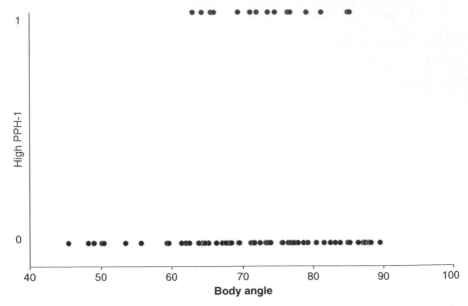

Figure 11.8   Scatter plot of HighPPH-1 versus body angle.

Binary Logistic Regression: HighPPH-1 versus Body Angle

```
Link Function: Logit


Response Information

Variable    Value   Count
HighPPH-1   1          16   (Event)
            0          66
            Total      82


Logistic Regression Table

                                                    Odds      95% CI
Predictor         Coef    SE Coef      Z      P    Ratio  Lower  Upper
Constant      -2.88561    2.01287  -1.43  0.152
Body Angle   0.0202122  0.0271819   0.74  0.457   1.02   0.97   1.08


Log-Likelihood = -40.189
Test that all slopes are zero: G = 0.567, DF = 1, P-Value = 0.451
```

Figure 11.9   Binary logistic regression results for HighPPH-1 versus body angle.

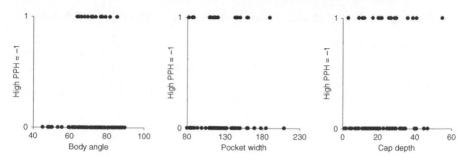

Figure 11.10   Scatter plots for HighPPH-1 versus continuous inputs.

model. Several techniques may be employed to select variables to include in a final logistic model.

We may initially look at variables one at a time to screen out those factors which clearly do not have a relationship. For continuous input variables, we recommend creating scatter plots. Figure 11.10 shows scatter plots for three continuous inputs: body angle, cap width, and pocket width. Significant factors typically will have an offset between their spreads for each binary response. Of these, we may expect that body angle would more likely appear in a final prediction model. Both pocket width and cap depth appear to have the same spread, regardless of the binary response. Caution is recommended with this approach. When multiple factors are changing simultaneously, the effect of a significant variable may be masked by variability caused by other factors.

Alternatively, we may use ordinary least squares regression methods to narrow the variables. Here, we recommend using best subsets regression or stepwise regression to narrow the list of potential variables. While these techniques have limitations in creating a final prediction model for a binary output, they often are effective at filtering out nonsignificant variables. Using these techniques, we may create the following prediction model shown in Figure 11.11. Of note, while *body angle* was not significant by itself, it is significant at an α level of 10% in the final model when also including the *feature line* variable.

With this final model, we should point out that the odds ratio is much larger in magnitude for the categorical input than the continuous variable; however, the width of the 95% confidence interval for the odds ratio also is much larger for the categorical variable. This result is reasonable as we expect a more consistent estimate for a continuous input with numerous levels versus a categorical variable with only a few distinct levels.

11.3   Logistic regression and customer loss functions

While identifying significant input variables is important, design engineers often prefer to understand those features and settings that may result in customer dissatisfaction. We seek to identify robust ranges for these settings to maximize design

Binary Logistic Regression: HighPPH-1 versus Feature Line, Body Angle

```
Link Function: Logit


Response Information

Variable   Value  Count
HighPPH-1  1         16   (Event)
           0         66
           Total     82


Logistic Regression Table

                                                     Odds      95% CI
Predictor          Coef     SE Coef      Z      P   Ratio  Lower  Upper
Constant        -6.23312    2.73358  -2.28  0.023
Feature Line     1.78942   0.737237   2.43  0.015   5.99   1.41  25.39
Body Angle      0.0584899  0.0347204   1.68  0.092   1.06   0.99   1.13


Log-Likelihood = -37.028
Test that all slopes are zero: G = 6.889, DF = 2, P-Value = 0.032
```

Figure 11.11   Final model for HighPPH-1.

flexibility and minimize cost. Here, binary logistic regression provides an effective method to create a loss function as the output variable may be expressed as the probability of an event occurring given a setting or range of settings for the explanatory input variables.

To illustrate, let us return to the fuel filler door example. Suppose the final model includes two significant explanatory input variables: *body angle* and *feature line*. Using binary logistic regression analysis, we may compute the event probabilities for all values for the inputs.

Alternatively, if we have the final model coefficients, we may estimate the probability using Equation 11.5. For instance, suppose the coefficients for β_0, β_1, and β_2 are −6.233, 0.058, and 1.789, respectively. For vehicle one, the *body angle* is 63 and the *feature line* coded result equals one. From here, we may use Equation 11.5 to estimate a probability of 0.3122.

For illustrative purposes, Table 11.3 shows sample results for ten of the vehicles in this study.

An effective graphical tool to show the relationship between the key explanatory input variables and the response for a binary logistic regression analysis is an event probability scatter plot (Figure 11.12). In this case, we have one continuous input, *body angle*, and another categorical input, *feature line*, which has two levels. As such, we may create an event probability plot with two curves, one curve for each level of the categorical input.

Table 11.3   Estimated event probabilities for the first 10 vehicles.

Vehicle ID	Constant	Body Angle	Feature Line	Probability of High PPH
1	−6.233	63.0	1	0.3122
2	−6.233	65.2	0	0.0793
3	−6.233	59.3	1	0.2680
4	−6.233	62.0	0	0.0668
5	−6.233	53.5	1	0.2073
6	−6.233	55.7	0	0.0473
7	−6.233	48.2	1	0.1613
8	−6.233	59.3	0	0.0577
9	−6.233	59.6	1	0.2714
10	−6.233	73.9	0	0.1249

The values displayed in this table have been computed using the rounded parameters shown in the text. Results will differ when software is used.

To make design recommendations, we must agree on a threshold of acceptance for the loss function. For a *things gone wrong* study, we may wish to identify settings or a range of settings yielding a dissatisfaction of not less than 15% of customers. We may wish to set a more stringent threshold of 5% or perhaps even 1%. Of course, setting

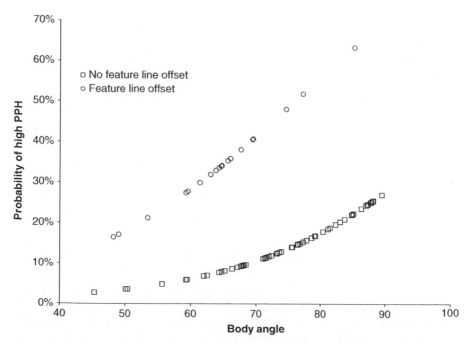

Figure 11.12   Event probability scatter plot.

a more stringent threshold may result in excessive cost or loss of design flexibility to satisfy relatively few customers. As an aside, for a *things gone right* study where higher percentages are sought, the threshold may be set at 85%, 95%, or 99%.

In this example, suppose we want to have a 5% or less likelihood of falling into the high PPH group. Here, we would recommend an angle of 57° or less with no body feature line offset through the fuel filler door.

While we may create a similar graph for any number of variables, we maintain that event probability plots are most appealing for logistic models with either one or two continuous input variables, or one continuous input variable with one or two categorical factors. In first case, we may create a three-dimensional scatter plot with the event probability on the Z axis and the numerical measures on the X axis and Y axis. In the latter case, we will have a curve for each combination of settings for the categorical input. Unless one has a very large sample size, we find that four is typically a reasonable limit for displaying results.

11.4 Loss function with maximum (or minimum) response

In the above study, we were able to assign a binary response for each sample vehicle. While this is the more common approach, we may wish to conduct a study where each participant is asked to indicate a maximum or minimum limit for usability or comfort. For instance, a vehicle manufacturer may wish to determine the maximum distance a customer would be willing to reach to perform a task. In another example, a manufacturer may wish to know how far away from a vehicle a driver may be before they want their doors automatically locked as a safety precaution.

To answer these types of questions, one could give each participant in the study an increasing distance and ask whether the participant would find the distance acceptable. Here, we could use the approach discussed in the prior section. The limitations of this approach, however, are that participants would need to make evaluations at numerous intervals increasing the time and cost of the study. In addition, we may not know the exact distance value at which a participant indicates an unacceptable condition. Instead of an equally spaced interval approach, we could modify the experiment and simply ask participants for a maximum acceptable distance.

To illustrate, an automotive manufacturer is interested in the reach distance for an instrument panel feature. The manufacturer brought in 32 representative customers to evaluate how far customers were willing to reach for a feature until they considered it objectionable as well as their preferred reach. For the study, the seat was positioned to the full down and full rear location prior to participants entering the vehicle. Participants were asked to sit in the vehicle, adjust the seat to a comfortable position, and then reach for a feature. This distance was recorded as their preferred location. Next, the participants were asked to re-position themselves just before they would be dissatisfied with the reach. This distance was recorded as the forward limit position.

To apply a binary logistic regression model to this example, one first has to make some assumptions. For instance, we will assume that if a participant indicates a

Table 11.4   Forward limit and preferred location measurements.

Forward Limit	Number Acceptable	Number Unacceptable
207	31	1
226	30	2
257	29	3
263	28	4
285	27	5
294	26	6
295	25	7
298	24	8
300	23	9
303	22	10
305	21	11
310	20	12
317	19	13
322	18	14
326	17	15
334	16	16
338	15	17
339	14	18
343	13	19
347	12	20
355	10	22
355	10	22
356	9	23
360	8	24
370	7	25
383	6	26
395	5	27
400	2	30
400	2	30
400	2	30
405	1	31
458	0	32

forward reach limit of 300 mm as its maximum point, then all participants having a limit less than or equal to 300 mm would find this reach unacceptable (code = 1), while all those having a limit greater than 300 mm would find it acceptable (code = 0). Thus, for each limit value, we will identify the number of participants who find it acceptable versus unacceptable. Table 11.4 summarizes the reach limit data for each value.

To use the binary logistic regression, we need to identify the response column (values of 0 and 1), the frequency column (frequency count for each row), and a model input variable, which would be the *forward limit* for this example. Table 11.5

Table 11.5   Forward limit data formatted for statistical analysis (first four rows).

Forward Limit	Count	Response
207	31	0
207	1	1
226	30	0
226	2	1

Binary Logistic Regression: Response versus Forward Limit

```
Link Function: Logit


Response Information

Variable  Value  Count
Response  1        532   (Event)
          0        492
          Total   1024

Frequency: Count

* NOTE * 63 cases were used
* NOTE * 1 cases contained missing values or was a case with zero frequency.


Logistic Regression Table

                                                   Odds     95% CI
Predictor            Coef    SE Coef      Z      P Ratio  Lower  Upper
Constant         -10.8131  0.726109  -14.89  0.000
Forward Limit  0.0328063  0.0021787   15.06  0.000  1.03   1.03   1.04
```

Figure 11.13   Binary logistic regression results for forward limit.

displays the first four rows of the data in Table 11.4 transformed into a format suitable for analysis.[3]

The binary logistic regression results are shown in Figure 11.13.

For this example, the final model coefficients for β_0 and β_1 (forward limit) are -10.8131 and 0.0328, respectively. As before, we may estimate an event probability for each input value in the study. With these, we then may create the event probability plot shown in Figure 11.14.

For this study, the dissatisfaction allowance threshold is 15% of customers. Therefore, we may recommend a reach limit value that does not exceed 277.

---

[3] Tables 11.4 and 11.5 (all 64 rows) are available on the accompanying website. The file name is "Tables_11_4_5."

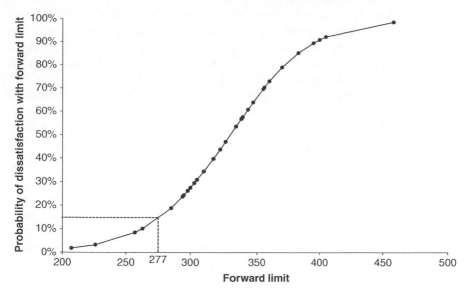

Figure 11.14   Event probability plot for forward limit.

11.5   Summary

Regression analysis does not require a continuous output. Many examples exist where the response is binary (or may be converted to a binary outcome), and thus we may apply binary logistic regression. With a binary logistic regression analysis, we may estimate coefficients and use a prediction equation to estimate the conditional probability for a binary event. We may then use this prediction equation to generate a loss function and display results using an event probability plot. From here, we may identify a setting or range of settings to minimize the likelihood of customer dissatisfaction.

Exercises

The data required to complete the following exercises are provided on the accompanying website. The file name is "Chapter11_Exercise_Data.xlsx."

1 A DVD distribution center has been experiencing complaints regarding software issues related to their order inquiry system. To examine the issue, they randomly select 400 recent calls where they obtain the information related to software glitches during searches. Their response is whether a search issue has occurred (yes = 1). For each inquiry, they also record the number of inquiries searched during a call. Given these data, run a binary logistic regression to determine the odds ratio between number of inquiries, and whether a search issue has occurred. Estimate the number of inquiries required to observe a search issue with a probability of 0.9?

2 A stamping manufacturer is experiencing high scrap rate due to splits in one of their main products. To address the issue, they conduct a study to assess the relationship between various material and process input variables on the output variable (either a split occurs or does not occur). Ninety parts are sampled under different factor settings and the total number of splits is 12.

Conduct a binary logistic regression analysis using the variables: amount of lubrication (ordinal variable) and blank holding force. Which of the variables has a statistically significant effect on the output split ($\alpha = 0.05$). If the goal is to minimize splits, make a recommendation based on the available information.

3 A manufacturer wishes to establish a relationship between vehicle size (size is measured by the product of vehicle width and height), maximum trailer pull weight, and whether they should use a unibody (code = 1) versus body-on-frame (code = 0) body design architecture. They obtain information on 41 vehicles. Run separate binary logistic regression for each input factor, vehicle size, and max trailer weight. Then, create an event probability plot between each of these inputs and type of body design (body-on-frame versus unibody). Provide estimates for both maximum trailer weight and vehicle width equating to a probability of 0.8 for selecting a unibody design.

12

Verification and validation

12.1 Introduction

Throughout this book, we have discussed techniques to develop optimal, robust products. In some cases, the primary focus has been on optimization techniques relying on engineering models, although we have also discussed using designed experiments and logistic regression analysis. Regardless of the approach, any design optimization effort requires confirmation.

As done in software engineering (IEEE, 1998), we may make a distinction between verification and validation. *Verification* involves the review and evaluation of designs, documents, and plans to ensure we have fulfilled the specified design requirements. In other words, *"Does the design satisfy the specifications?"* Usually, this is done prior to physical evaluation using actual production intent processes. It may be performed using design checklists, walkthroughs, meetings, and simulations. Typically, it is a virtual or *paper* exercise.

Validation, on the other hand, involves the review and physical evaluation of the actual product. It involves physical testing of the product relative to customer and functional requirements and requires acceptance or approval by the next process customer or end user. It answers the question, *"Does the design satisfy the customer?"* In some cases, final validation may not be completed until after a product has been produced using the actual production process and is in use by customers.

As per the above operational definitions, verification should take place prior to validation and not vice versa. Still, verification and validation may be viewed as a continuous evaluation process during new product development (NPD), where the results of each assessment helps one converge on a desirable end product or service.

Probabilistic Design for Optimization and Robustness for Engineers, First Edition.
Bryan Dodson, Patrick C. Hammett and René Klerx.
© 2014 John Wiley & Sons, Ltd. Published 2014 by John Wiley & Sons, Ltd.
Companion website: http://www.wiley.com/go/robustness_for_engineers

Verification and validation together ensure a design:

- meets all customer requirements,

- meets all functional and design parameter requirements while being robust to expected usage conditions,

- meets all functional and design parameter requirements for the required product life, and

- may be produced consistently without special controls or inspections.

Figure 12.1 displays a variation of the systems engineering V-diagram described in Chapter 1. In systems engineering, verification and validation (V&V) of the evolving project documentation is performed early to maximize the probability of identifying defects as early as possible in the project development cycle. V&V begins at the lowest level of the system and evolves up to the system level. Particular attention is paid to those design areas with lack of knowledge, assumptions, and high sensitivity. Sources of variation are another primary target of a V&V plan, as variability in use and manufacturing are often causes of failure.

As mentioned in Chapter 1, boundary diagrams may be used to define and cascade requirements from the system to lower levels of responsibility. When these requirements are cascaded down, more detail is revealed, and areas with lack of knowledge are identified as well as where assumptions have been made.

Consider the projectile distance problem described in earlier chapters. At the system level, if the angle and velocity requirements are met, the distance requirement is satisfied. Cascading the velocity requirement down, other parameters become important, such as the weight of the projectile and the friction between the barrel wall and the projectile. Costly and time-consuming changes are inevitable if specifications for projectile weight and friction are not determined quickly in the project timeline. The friction is an area of uncertainty, especially since the same organization may not have responsibility for the barrel wall and the projectile. As opposed to validating the entire velocity system, V&V should be targeted to areas where assumptions have been made, or where there is a lack of knowledge. In this case, the lack of knowledge is the friction between the barrel wall and the projectile.

When creating engineering designs, models are used to predict behavior. Some models are exact, while others only provide rough estimates. Any assumptions in models should be documented, and eventually become a target in the V&V plan. Consider the brick wall from Chapter 9. The engineering model seems simple and accurate; the height of the wall is the sum of the heights of the individual bricks. Closer scrutiny reveals two assumptions in the model, the bricks are assumed to be flat, and temperature is assumed to have no effect on height. In this case, the V&V plan may assess the magnitude of these assumptions. If the magnitude is small enough, the system C_{pk} will be reduced from 1.67, but will still be acceptable. If the magnitude is too large, flatness or temperature will have to be included in the height model or accounted for in some other manner.

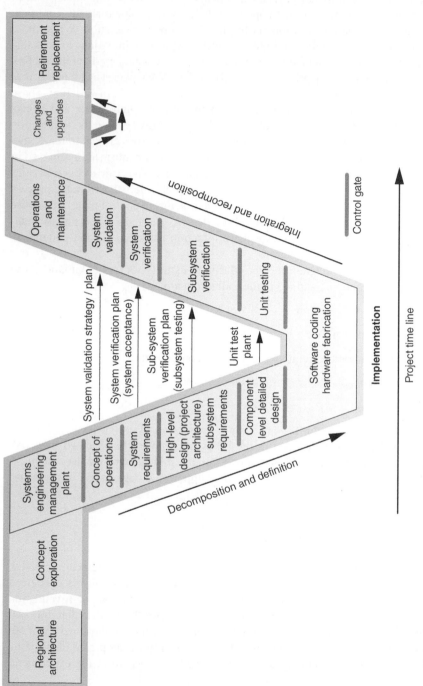

Figure 12.1 System engineering V-diagram.[1]

[1]Taken from the US Department of Transportation, publication No. FHWA-HOP-07-069, January 2007.

The robust design model provides a sensitivity analysis for all inputs. Recall the reservoir flow problem from Chapter 7. The flow rate, pipe diameter, and pipe friction factor are the top contributors to pump horsepower variation. While these factors should be validated during model development, it is important to pass this information forward to manufacturing or purchasing. The boundary diagram is a link in a chain of documents that provide the backbone for a V&V plan. The boundary diagram is augmented with a parameter diagram, which lists sources of variation. These sources of variation become failure causes in the design failure mode and effects analysis (FMEA). The design controls in the design FMEA become the verification plan. The design FMEA also identifies potential characteristics critical to the design function and to safety. The critical characteristics are either confirmed or removed in the process FMEA based on the manufacturing capability. The critical characteristics information form summarizes the confirmed critical characteristics from the process FMEA, and becomes the foundation for the manufacturing control plan. This entire chain of documents exists at each level of the design. These documents allow manufacturing and purchasing to validate the production processes for the parameters most critical to design system outputs.

In Chapter 3, the P-diagram was introduced to identify sources of variation. The five sources of variation identified in the P-diagram are:

1. piece-to-piece variation,

2. degradation over time,

3. interactions with surrounding systems or components,

4. environment, and

5. customer use.

The sources of variation become failure causes in the design FMEA. Sources of variation that are not included in the robustness model, but deemed high risk should be assessed in the V&V plan. For example, a P-diagram for the distance a projectile travels may include stiffness of the support and degradation of the barrel surface finish.

12.2   Engineering model V&V[2]

Figure 12.2 depicts a process for verification and validation of an engineering model.

On the experimental side of Figure 12.2, a physical experiment is conceived and designed. The result is a validation experiment. The purpose of a validation experiment is to provide information needed to validate the model; therefore, all assumptions must be understood, well defined, and controlled in the experiment. To assist with this process, calculations may be performed, for example, to identify the most effective locations and types of measurements needed from the experiment.

---

[2]Taken from the Los Alamos National Laboratory, publication LA-14167-MS, October 2004.

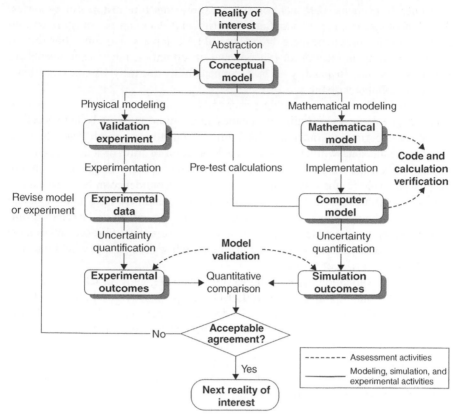

Figure 12.2   Engineering model V&V process.[3]

These data will include not only response measurements, but also measurements needed to define model inputs and model input uncertainties associated with loadings, initial conditions, boundary conditions, etc. For example, load and material variabilities can be quantified by the symmetrical placement of sensors within an experiment, and test-to-test variations can be quantified by performing multiple validation experiments.

The pre-test calculations link shown in Figure 12.2 also reflects the important interaction between the designer and the experimenter that must occur to ensure that the measured data are needed, relevant, and accurate. Once the validation experiment and pre-test calculations are completed, however, the modeler and experimenter should work independently until reaching the point of comparing outcomes from the experiment and the simulation.

Experimentation involves the collection of raw data from the various sensors used in the physical experiment (strain and pressure gauges, high-speed photography, etc.)

---

[3]Taken from the Los Alamos National Laboratory, publication LA-14167-MS, October 2004.

to produce experimental data such as strain measurements, time histories of responses, videos, photographs, etc. If necessary, experimental data can be transformed into experimental features to be more directly useful for comparison to simulation results. To support the quantification of experimental uncertainties, repeat experiments are generally necessary to quantify the lack of repeatability due to systematic error (bias) and uncontrollable variability.

Uncertainty quantification is then performed to quantify the effect of measurement error, design tolerances, as-built uncertainties, fabrication errors, and other uncertainties on the experimental outcomes. Experimental outcomes typically take the form of experimental data with error bounds as a function of time or load. Uncertainty quantification is shown on both left and right branches of the V-diagram to highlight its important role in quantifying the uncertainty and confidence in both the experimental and simulation outcomes. The quantitative comparison of experimental and simulation outcomes may take the form of a statistical statement of the selected validation metrics. For example, if the validation metric were the difference between the simulation and experimental outcome, the quantitative comparison would quantify the expected accuracy of the model.

The model validation assessment determines the degree to which a model is an accurate representation of the real world from the perspective of the intended uses of the model. This information is used to decide whether the model has resulted in acceptable agreement with the experiment. The question of whether the model is adequate for its intended use is broader than implied in the acceptable agreement decision block shown in Figure 12.2. The acceptable agreement decision focuses only on the level of agreement between experimental and simulation outcomes, the criteria for which were specified as part of the conceptual model.

12.3   Design verification methods and tools

Design verification typically begins in the front end of the design process, utilizing methods ranging from design reviews to physical testing. Among some commonly used tools and methods are:

- design verification reviews,
- virtual manufacturing and assembly,
- prototypes and early production builds,
- confirmation tests comparing alternatives, and
- confirmation tests comparing the design to acceptance criteria.

12.3.1   Design verification reviews

Organizations have design review processes to identify potential product design-related issues and operational concerns. Further, many organizations use cross-functional teams (designers, process engineers, and manufacturing personnel) to conduct these reviews to promote simultaneous or concurrent engineering (Nevins et al., 1989; Clark & Fujimoto, 1991).

While many organizations use cross-functional design reviews, the degree of integration, authority, and expertise levels among resources may vary significantly across organizations (or even within a company). As discussed in Chapter 1, we maintain the most effective design verification reviews allow for downstream functions, such as manufacturing engineers, to have some authority in the acceptance of a design and perhaps more importantly, have rights to reject or require changes to a design proposal. To do so, designers must release sufficient information to allow downstream users to identify potential concerns and make informed decisions. In addition, organizations must assign the right downstream process resources with the necessary knowledge and manufacturing expertise to identify potential concerns to be included in process validation activities in the back end of the NPD.

For effective design reviews, organizations first must establish requirements and criteria for their key product and process subsystems at project milestones or phases within the NPD process. Then, organizations must hold resources accountable to meet these requirements and criteria. Today, these requirements and criteria often are communicated and monitored using design checklists, and the design documents listed earlier (boundary diagram, P-diagram, design FMEA, process FMEA). A simple, common method to monitor progress and summarize conformance is through status color designations. For instance, a *red*, *yellow*, *green* design status review report may look as follows:

- *Green*—subsystem has passed all the milestone gate criteria with no major problems and is ready for the next activities.

- *Yellow*—subsystem has passed some milestone gate criteria, but has minor problems that must be corrected for passage. This typically requires the development of a recovery plan rather than a stop in the design process.

- *Red*—subsystem has passed some gate criteria, but has one or more major problems that must be addressed before proceeding.

Of course, the operational meaning of green, yellow, red, or any other color designation ultimately must be set by an organization to reflect its requirements to produce desirable products.

A common approach used to incorporate reviews and move the design from concept to launch is a gate review process. Cooper (2011) provides a comprehensive discussion of a stage-gate process.

12.3.2   Virtual prototypes and simulation

To verify product design concepts and intent prior to physical validations, many organizations perform virtual studies or simulations including virtual manufacturing and assembly validation. Virtual prototypes involve using computer-aided design (CAD) and other computer engineering tools to validate the design before physical parts are constructed. This may be done by generating two-dimensional and three-dimensional computer-developed shapes into parts and then combining or assembling them virtually to identify potential design concerns. When done properly, virtual prototyping is more than just an aesthetic review, but allows designers to explore fit and functional-related issues and go through problem-solving cycles faster and at lower cost through front-loading problem solving (Thomke & Fujimoto, 2000).

12.3.3   Physical prototypes and early production builds

Functional prototypes provide a replica or model to allow engineers to evaluate design intent, identify potential problems, and test problem resolutions. For practical reasons, functional prototypes often are characterized by:

- use of non-production intent materials,
- use of non-production processes or operations,
- allowance for greater deviations from design intent than in actual production.

Similar to functional prototypes are early production builds. The main difference is that these typically use production intent material and often the production intent process. For example, a manufacturing unit may produce a stamped part using a subset of the regular production die operations but then manually perform some operation using more general purpose equipment such as lasers for trimming and cutting sheet metal to create an actual part.

Both functional prototypes and early production builds are useful in identifying potential downstream manufacturability issues. They also may be used to initially verify or confirm some design requirements. Of course, organizations must verify and validate later using actual production processes and materials.

12.3.4   Confirmation testing comparing alternatives

For design verification, tests may be conducted using virtual simulations, prototypes, or early production builds to confirm that a proposed design meets its requirements and ideally provides a robust, optimal solution. If not, organizations should at least verify it is better than its alternatives, until a new design solution is developed, possibly in a subsequent product release.

To compare a new product design against a set of alternatives, one must first establish the metrics and criteria to make an assessment. A product design scorecard provides an excellent tool to summarize these metrics including target values, acceptance limits, and current performance levels to measure a design's overall desirability.

Comparing alternatives statistically is difficult because of the difficulty of obtaining a random sample representative of the population. If a random sample is not used, any statistical test is invalid. The following sources of variation are often missing from prototypes:

- variation from multiple batches of raw material,
- tooling changes,

- equipment maintenance, and

- multiple operators.

12.3.5 Confirmation tests comparing the design to acceptance criteria

Once a design solution is selected (ideally an optimal robust one), product developers may run confirmation tests to verify the design against areas of uncertainty, model assumptions, or sources of noise. A design scorecard assessment is an effective way to display and summarize results of a functional audit or performance test. Scorecards should include key requirements and their acceptance criteria. During design verification and validation (virtual or physical) and later during process validation, organizations may run tests to compare actual performance to design criteria. They may summarize the overall conformance using the total desirability index.

Another common validation technique to determine the expected life of a product to function as intended is a reliability analysis or assessment. Reliability assessments often are done prior to manufacturing process validation to verify that a design is robust to its expected user conditions and meets its design life objectives for product warranty. Reliability assessments typically involve exposing the product to likely uncontrollable environmental noise factors and then evaluating how well it performs under such conditions. A common reliability measure is the time until failure. Given failure times from a reliability assessment, one may estimate the expected design life. Moreover, a reliability test may also identify which components are likely to fail first and under what specific operating conditions.

Reliability tests may be completed using field tests or in a laboratory environment. For design verification and validation, a laboratory or test bed is often used as it provides a faster and less costly way to identify potential failure modes and predict the design life. Still, field studies should be incorporated where possible to test for unforeseen user conditions or reflect conditions that are hard to replicate in a laboratory environment. For complex products, manufacturers should create a reliability test plan to ensure potential failure modes are evaluated and that the final product meets all its performance objectives in terms of its design life.

12.4 Process validation procedure

Process validation activities typically occur during the back end of product development using build phases often referred to as tool tryout, manufacturing validation, and launch. These validation activities are similar, and in some cases the same, as the tasks used to approve parts for the start of production.

The steps below are for mass production. Many engineering designs have little or no need for process validation. For example, a bridge will only be produced once, and many other products have a production rate of less than ten per year.

There are four key steps in process validation. First, we must fulfill all production readiness requirements. This includes verifying all final validation criteria are established with clearly defined acceptance criteria, processes have defined standardized work procedures, and capable measurement systems are in place to assess functional, design, and process requirements. Next, we may conduct static process validation to ensure processes work as intended and are capable of producing acceptable products, though not necessarily while being manufactured at desired production cycle times or full production line rates. Then, we may perform dynamic process validation. We again validate that the process is capable of producing acceptable products, but at full production rate. Finally, we have the production approval process. Here, we validate that processes are producing all product outputs as per their design specifications, at desired quality levels, at full production rate, and with a control plan in place to sustain long-term stability and process control. We now discuss each of these steps further in detail.

Assessing production readiness is the first step in effective process validation. This may be accomplished by verifying all design and process planning documents needed to perform validation are complete per a validation requirement checklist (Table 12.1). Readiness documents may include a design status review checklist, an FMEA, a list of key product and process requirements and specifications, standard operating procedures, training requirements checklist, and a readiness assessment for the measurement system.

Table 12.1   Sample production readiness checklist.

Item	Description
1	Is the product design FMEA complete? Have all high-risk issues been resolved?
2	Is the design scorecard updated to latest level?
3	Have all product functional and design requirements been verified (meet all performance, usability, and reliability requirements)?
4	Does the production facility have adequate capacity to meet customer demand?
5	Is the supporting packaging, handling, storage, and shipping at the production facility properly planned?
6	Have all supplier processes and tooling been verified?
7	Is the supplier capability adequate for effective process validation of the product?
8	Is the production tooling and any special test equipment ready for production?
9	Are adequate measurements systems in place to measure product performance?
10	Have all operator work tasks been reviewed and validated for safety and ergonomic requirements?
11	Have all standardized work instructions been reviewed and validated?
12	Have appropriate training requirements been established and documented?

A readiness assessment should begin with a status check on design completion. For instance, organizations should ensure all product drawings are up-to-date, all major failure modes and design risks have been identified and resolved, and all major design changes have been incorporated into both the product and its related process tooling. In those cases where key changes may be pending, organizations should have a recovery plan in place. Beginning static and dynamic validation without a stable design is usually an exercise in futility as subsequent validation steps will need to be repeated after changes are made.

In addition to a stable design, organizations should ensure that suppliers have their processes validated and are capable of supplying quality components. Similarly, a production readiness assessment needs to verify that all key supporting processes are in place. For example, organizations need to have a place to store incoming components and finished products before starting process validation. Sometimes this requires making special packaging or racks to store products without damage.

Next, organizations should confirm that all processes have been reviewed and validated per safety and ergonomic requirements and that standardized work procedures have been developed. Operators should have a clear understanding of the required process outputs, any potential safety-related concerns, and the expected process completion time and quality performance requirements prior to starting process validation. For example, instructions should be provided on how to perform each task, and adequate training should be given on how to use and maintain any required equipment. All these should be communicated as visually as possible using tools such as process layout diagrams, step-by-step pictorial guides, and illustrations showing both acceptable and unacceptable products.

Standardized work packages, especially during the validation phase, are rarely, if ever, fully complete and self-sustaining. They require periodic reviews and audits. Thus, assessing preparedness during process validation is not a one-time event. It must be revisited and with procedures being revised as appropriate after each subsequent validation step. Standardized work procedures should be revisited again after starting regular production as part of continuous process improvement efforts.

In addition, a production readiness assessment must verify that all test and inspection procedures are defined and accurate and capable measurement systems exist to measure the product characteristics. For instance, organizations should already have verified that all inspection gages and measurement equipment meet their accuracy requirements and pass gage repeatability and reproducibility requirements. Failure to do so prior to process validation can significantly hamper meeting process validation milestone reviews and potentially waste resources trying to resolve out-of-specification product characteristics whose root cause for their condition may be the measurement system itself.

Ultimately, all organizations must establish their own readiness assessment checklist and guidelines to meet the needs of their particular business. For an example of general guidelines, we refer readers to the Advanced Product Quality Planning (APQP) and Control Plan Reference Manual produced by the Automotive Industry Action Group (2008).

Once the production readiness checklist items have been verified, we may begin process validation. In doing so, we recommend validating processes in two stages: static and dynamic process validation. Static process validation is used to validate that a process operates as intended and meets all requirements without concern for production line rate. In contrast, dynamic process validation is used to validate that a process operates as intended and meets all requirements at full production line rate.

The primary reason for separating these activities is that running operations at full production line rate often involves additional potential sources of variation that must be considered. As a classic simple example, the process of making a single pizza is different than the process of making thousands of pizzas. At full line rate, we have to be concerned that the automation is functioning properly, ergonomic issues are resolved, wasted motion is minimized, and all tools, equipment, and material supplies are easily accessible. As such, we first recommend performing a static process validation where the focus is on validating the process tooling and equipment are built correctly, the measurement system is accurate and capable, and the process has the potential to produce products that meet specifications. In fact, the high capability a process often may achieve is during static mode where great care is put into the loading, cycling, and handling of parts during a processing operation.

Some examples of static process validation activities include:

- tooling certification studies confirming all process tooling is built and tuned— as per the tool design specifications,

- measurement systems analysis studies confirming all measurement systems are accurate and meet repeatability and reproducibility requirements,

- static mode functional testing confirming the processes are functioning properly prior to adding automation or running at full line rate, and

- static process capability studies confirming the process is capable of producing products within specification.

If process tooling and equipment are built at external suppliers, static validation may be included in supplier tooling validation and approval studies. In such instances, the above activities typically need to be repeated once tooling and equipment are installed at the home production facility.

After meeting expectations for static process validation, we may focus on dynamic, run-at-rate, production trials. Here, organizations validate that the process functions as intended at full production line rate. Dynamic validation also should include process capability studies to validate that the process is capable of producing a product that meets all performance requirements and specifications while running at desired process cycle time. Such studies typically involve measuring larger sample sizes of 30-100, with samples spread out over multiple setups. These studies typically include both an assessment of stability using an appropriate process control chart and a process capability study. Together these studies often are incorporated into a formal Production Part Approval Process (PPAP).

Table 12.2   AIAG PPAP checklist.

Item	Description
1	*Design record.* Are all requirements for the design record met (e.g., CAD/CAM math data available, part drawings, geometric dimensioning, and tolerance drawings where specified)?
2	*Engineering change documents.* Are any engineering changes incorporated in the design record not in the product?
3	*Design FMEA.* Has a design FMEA been completed and all high-risk items identified and resolved?
4	*Process flow diagram and standardized work document.* Has the process flow been clearly identified along with detailed instructions for operators to perform the process?
5	*Process FMEA.* Has a process FMEA been completed and all high-risk items identified and resolved?
6	*Product design functionality, performance, and reliability.* Does the product meet all its functional, performance, and reliability requirements (e.g., reliability test plan)?
7	*Measurement systems.* Have all measurement systems used for process stability and capability studies been validated to meet requirements for accuracy and capability (e.g., gage R&R)?
8	*Dimensional results of samples.* Have part samples from the actual production process been checked against their specifications with results clearly communicated on part inspection reports?
9	*Initial process stability.* Has the process been deemed stable or in statistical control using appropriate assessment methods (e.g., trend or statistical process control (SPC) charts)?
10	*Initial process capability.* Do all key product characteristics meet their requirements for process capability (e.g., $P_{pk} > 1.33$ or meet % in specification requirement)?
11	*Cosmetic validation.* Have all cosmetic and aesthetic requirements by the customer been satisfied?
12	*Control plan.* Has an adequate control plan for all product and process-related characteristics been established and verified?

Many organizations develop specific part approval processes with criteria that must be met prior to the start of regular production. Table 12.2 provides a sample checklist of approval elements based on the North American automotive manufacturers PPAP.[4] These elements include documents, measurement systems analysis, process stability analysis, and process capability studies.

---

[4] Production Part Approval Process is described in detail in AIAG PPAP-4, 4th edition, 2006.

Although we have presented the above process validation steps in a sequential manner, they may involve a series of tests or iterations within each along with modifications and re-work required to pass validation requirements. As such, effective process validation clearly involves monitoring progress throughout the NPD. Monitoring may range from the usage of issues tracking lists to progress monitoring charts for various metrics such as:

- percentage of issues resolved,

- total product design and engineering costs,

- percentage of final product dimensions within specification, and

- percentage of sub-components approved for production.

Again, the actual metrics will vary by manufacturer. Still, organization should seek a balance of leading and lagging indicators. For instance, tracking percentage completion to an upcoming milestone event (leading indicator) is often more useful than recording the number of late components for a process validation build event (lagging indicator). In the former case, a manufacturer has a chance to recover, whereas in the second case they may only offer an apology.

12.5   Summary

Throughout this book, we have presented several methodologies from engineering models to designed experiments to develop optimal, robust products. Regardless of the approach, all conclusions and recommendations will require some form of validation or confirmation study. Such a validation activity, however, is only one element within a set of design verification and process validation activities needed to occur throughout the NPD process. In this chapter, we provided an overview of various tools and methods. These include design verification methods such as design reviews, simulations, prototypes, and confirmation tests. After the design is stable, organizations may begin process validation which involves production readiness assessments, static and dynamic process validation, and part approval processes to ultimately ensure products meet performance and usability needs of customers.

References

Agresti, A. (1996). *An introduction to categorical data analysis*. New York, NY: Wiley.

Assistant Secretary of Defense for Research and Engineering. (2011). *Technology Readiness Assessment (TRA) Guidance*. Washington, DC: Department of Defense.

Automotive Industry Action Group. (2006). *Production part approval process (PPAP)* (4th ed.). Southfleld, MI: AIAG.

Automotive Industry Action Group. (2008). *Advanced product quality planning (APQP) and control plan: Reference manual* (2nd ed.). Southfleld, MI: AIAG.

Automotive Industry Action Group. (2008). *Potential failure mode and effects analysis (FMEA): Reference manual* (4th ed.). Southfleld, MI: AIAG.

Berger, C., Blauth, R., Boger, D., Bolster, C., Burchill, G., DuMouchel, W., Pouliot,F., Richter, R., Rubinoff, A., Shen, D., Timko, M., & Walden, D. (1993). Kano's methods for understanding customer-defined quality. *Center for Quality of Management Journal*, *2*(4), 3–36.

Bothe, D. R. (1997). *Measuring process capability: Techniques and calculations for quality and manufacturing engineers*. New York, NY: McGraw-Hill.

Clark, K. B., & Fujimoto, T. (1991). *Product development performance: Strategy, organization, and management in the world auto industry*. Boston, MA: Harvard Business School Press.

Clausing, D. P. (1994). *Total quality development: A step-by-step guide to world class concurrent engineering*. New York, NY: ASME Press.

Cooper, R. G. (2011). *Winning at new products: Creating value through innovation*. New York, NY: Basic Books.

Costa, M. J. (2005). Applying Six Sigma to business process excellence. *BP Trends*, 1–5, January 4, 2005.

Creveling, C. M., Slutsky, J., & Antis, D. (2003). *Design for Six Sigma in technology and product development*. Upper Saddle River, NJ: Prentice Hall.

Derringer, G. C. (1994). A balancing act: Optimizing a product's properties. *Quality Progress*, (6), 51–58.

Derringer, G. C., & Suich, R. (1980). Simultaneous optimization of several response variables. *Journal of Quality Technology*, *12*(4), 214–219.

Probabilistic Design for Optimization and Robustness for Engineers, First Edition.
Bryan Dodson, Patrick C. Hammett and René Klerx.
© 2014 John Wiley & Sons, Ltd. Published 2014 by John Wiley & Sons, Ltd.
Companion website: http://www.wiley.com/go/robustness_for_engineers

Doane, D. P., & Seward, L. W. (2007). *Applied statistics in business and economics*. Boston, MA: McGraw-Hill.

Forsberg, K., & Mooz, H. (1991). *The relationship of systems engineering to the project cycle*. Center for Systems Management.

Harrington, E. C. Jr. (1965). The desirability function. *Industrial Quality Control, 21*(10), 494–498.

Heincke, M. (2006). *Leaking rear axles: A Design for Six Sigma project at General Motors* (2006:239 CIV). Luleå, Sweden: Luleå University of Technology.

Hilbe, J. M. (2009). *Logistic regression models*. Boca Raton, FL: CRC Press.

Hutcheson, G., & Sofroniou, N. (1999). *The multivariate social scientist: Introductory statistics using generalized linear models*. London, UK: Sage Publications.

IEEE Computer Society. (1998). *1012 - IEEE standard for software verification and validation; IEEE Std 1012-2004*. New York, NY: Institute of Electrical and Electronics Engineers.

Kasser, J. E. (1995). *Applying total quality management to systems engineering*. Boston, MA: Artech House.

Kasser, J., Frank, M., & Zhao, Y. Y. (2010) Assessing the competencies of systems engineers. In Proceedings of the 7th Bi-Annual European Systems Engineering Conference (EUSEC), Stockholm, Sweden, May 23–26, 2010.

Khuri, A. I., & Conlon, M. (1981). Simultaneous optimization of multiple responses represented by polynomial regression functions. *Technometrics, 23*(4), 363–375.

Kim, K.J., & Lin, J. (1999). Simultaneous optimization of mechanical properties of steel by maximizing desirability functions. *Applied Statistics, 49*(3), 311–326.

Kim, K. J., & Lin, D. K. (2000). Simultaneous optimization of mechanical properties of steel by maximizing exponential desirability functions. *Journal of the Royal Statistical Society Series C - Applied Statistics, 49*(3), 311–326.

Liggett, J. V. (1993). *Dimensional variation management handbook - A guide for quality, design, and manufacturing engineers*. Englewood Cliffs, NJ: Prentice Hall.

Long, J. S. (1997). *Regression models for categorical and limited dependent variables*. Thousand Oaks, CA: Sage Publications.

Mader, D. P. (2003). DfSS and your current design process. *Quality Progress, 36*(7), 88–89.

Mavris, D. N., & DeLaurentis, D. A. (2000). *Methodology for examining the simultaneous impact of requirements, vehicle characteristics, and technologies on military aircraft design*. Presented at the 22nd Congress of the International Council on the Aeronautical Sciences (ICAS), Harrogate, England, August 27–31, 2000.

Menard, S.W. (2002). Applied Logistic Regression Analysis. Los Angeles: SAGE.

Menard, S.W. (2010). Logistic regression: From introductory to advanced concepts and applications. Los Angeles: SAGE.

Montgomery, D. C., & Runger, G. C. (2011). *Applied statistics and probability for engineers* (5th ed.). Hoboken, NJ: Wiley.

Montgomery, D. C., Peck, E. A., Vining, G. G., & Ryan, A. G. (2012). *Student solutions manual to accompany: Introduction to linear regression analysis* (5th ed.). Hoboken, NJ: Wiley.

Morgan, J. M., & Liker, J. K. (2006). *The Toyota product development system: Integrating people, process, and technology*. New York, NY: Productivity Press.

Naumann, E., & Hoisington, S. H. (2001). *Customer centered Six Sigma: Linking customers, process improvement, and financial results.* Milwaukee, WI: ASQ Quality Press.

Nevins, J. L., Whitney, D. E., & De, F. T. (1989). *Concurrent design of products and processes: A strategy for the next generation in manufacturing.* New York, NY: McGraw-Hill.

Nielsen, J. (1993). Usability Engineering, Boston, MA.

Pohlman, J. T., & Leitner, D. W. (2003). A comparison of ordinary least squares and logistic regression. *Ohio Journal of Science, 103*(5), 118–125.

Ribardo, C., & Allen, T. T. (2003). An alternative desirability function for achieving "Six Sigma" quality. *Quality and Reliability Engineering International, 19*(3), 227–240.

Rook, J. (1986). *Communication for work: Written and oral assignments in communication.* Basingstoke: Macmillan Education.

Shewhart, W. A. (1931). *Economic control of quality of manufactured product.* New York, NY: D. Van Nostrand Company.

Smith, S. M., & Albaum, G. S. (2005). *Fundamentals of marketing research.* Thousand Oaks, CA: Sage Publications.

Snee, R. D., & Hoerl, R. W. (2003). *Leading Six Sigma: A step-by-step guide based on experience with GE and other Six Sigma companies.* Upper Saddle River, NJ: Prentice Hall.

Soderborg, N. R. (2004). Design for Six Sigma at Ford. *Six Sigma Forum Magazine, 4*(1), 15–22.

Stamatis, D. H. (2004). *Six Sigma fundamentals: A complete guide to the system, methods and tools.* New York, NY: Productivity Press.

Taguchi, G., & Clausing, D. (1990). Robust quality. *Harvard Business Review, 68*(1), 65–75.

Taguchi, G., Chowdhury, S., & Wu, Y. (2005). *Taguchi's quality engineering - Handbook.* Hoboken, NJ: Wiley.

Tennant, G. (2002). *Design for Six Sigma: Launching new products and services without failure.* Aldershot, UK: Gower.

Thomke, S. H., & Fujimoto, T. (2000). The effect of "front-loading" problem-solving on product development performance. *Journal of Product Innovation Management, 17*(2), 128–142.

Vanek, F., Jackson, P., & Grzybowski, R. (2008). Systems engineering metrics and applications in product development: A critical literature review and agenda for further research. *Systems Engineering, 11*(2), 107–124.

Ward, A., Liker, J. K., Cristiano, J. J., & Sobek II, D. K. (1995). The second Toyota paradox: How delaying decisions can make better cars faster. *Sloan Management Review, 36*(3), 43–62.

Yang, K., & El-Haik, B. (2003). *Design for Six Sigma: A roadmap for product development.* New York, NY: McGraw-Hill.

Bibliography

Balling, R. J., & Sobieszczanski-Sobieski, J. (1994). An algorithm for solving the system-level problem in multilevel optimization (AIAA-94-4330-CP). In M. F. Card (Ed.), *Proceedings of 5th AIAA/USAF/NASA/ISSMO symposium on "Multidisciplinary analysis and optimization", and held in Panama City, FL, September 7–9, 1994 (Part 2)* (pp. 794–809). Hampton, VA: NASA Langley Research Center.

Bloebaum, C. L., Hajela, P., & Sobieszczanski-Sobieski, J. (1992). Non-hierarchic system decomposition in structural optimization. *Engineering Optimization, 19*(3), 171–186.

Braun, R. D., Kroo, I. M., & Moore, A. A. (1996). Use of the collaborative optimization architecture for launch vehicle design (AIAA 96-4018). *Proceedings of 6th AIAA/USAF/NASA/ISSMO symposium on "Multidisciplinary analysis and optimization," and in Bellevue, WA, September 4–6, 1996 (Part 1)* (pp. 306–318). Washington, DC: American Institute of Aeronautics and Astronautics.

Broemm, W. J., Ellner, P. M., & Woodworth, W. J. (1999). *AMSAA reliability growth handbook (HDBK-A-1)*. Washington, DC: U.S. Army Materiel Systems Analysis Activity.

Castrup, H. (2009). *Error distribution variances and other statistics* (2009-01-15). Integrated Sciences Group.

Chan, H. A., & Englert, P. J. (Eds.). (2001). *Accelerated stress testing handbook: Guide for achieving quality products*. New York, NY: IEEE Press.

Chang, T. S., Ward, A. C., & Lee, J. (1994). Distributed design with conceptual robustness: A procedure based on Taguchi's parameter design. In R. Gadh (Ed.), *ASME concurrent product process design conference, Chicago, IL, 1994 (74,* pp. 19–29). New York, NY: ASME.

Chen, W., Allen, J. K., Tsui, K. L., & Mistree, F. (1996). A procedure for robust design: Minimizing variations caused by noise factors and control factors. *ASME Journal of Mechanical Design, 118*(4), 478–485.

De Levie, R. (2009). An improved numerical approximation for the first derivative. *Journal of Chemical Sciences, 121*(5), 935–950.

Dhillon, B. S. (1985). *Quality control, reliability, and engineering design*. New York, NY: Marcel Dekker.

Probabilistic Design for Optimization and Robustness for Engineers, First Edition.
Bryan Dodson, Patrick C. Hammett and René Klerx.
© 2014 John Wiley & Sons, Ltd. Published 2014 by John Wiley & Sons, Ltd.
Companion website: http://www.wiley.com/go/robustness_for_engineers

Dodson, B. L., & Klerx, R. (2012). An engineering alternative to statistically designed acceptance tests. *SAE International Journal of Passenger Cars—Mechanical Systems*, *5*(1), 161–166.

Dodson, B., & Nolan, D. (1999). *Reliability engineering handbook*. New York, NY: Marcel Dekker.

Dodson, B., & Schwab, H. (2006). *Accelerated testing: A practitioner's guide to accelerated and reliability testing*. Warrendale, PA: SAE International.

Dovich, R. A. (1990). *Reliability statistics*. Milwaukee, WI: ASQ Quality Press.

Duckworth, W. M., & Stephenson, W. R. (2000). Beyond traditional statistical methods. *The American Statistician*, *56*(3), 230–233.

Guilford, J., & Turner, J. U. (1993). Advanced analysis and synthesis for geometric tolerances. *Manufacturing Review*, *6*(4), 305–313.

Hammett, P. C., Hancock, W. M., & Baron, J. S. (1995). Producing a world-class automotive body. In J. K. Liker, J. E. Ettlie, & J. C. Campbell (Eds.), *Engineered in Japan: Japanese technology-management practices* (pp. 237–262). New York, NY: Oxford University Press.

Hammett, P. C., Wahl, S. M., & Baron, J. S. (1999). Using flexible criteria to improve manufacturing validation during product development. *Concurrent Engineering: Research and Applications*, *7*, 309–318.

Hobbs, G. K. (2000). *Accelerated reliability engineering: HALT and HASS*. Chichester, UK: Wiley.

Ireson, W. G., Coombs, Jr., C. F., & Moss, R. Y. (1996). *Handbook of reliability engineering and management* (2nd ed.). New York, NY: McGraw-Hill.

Kapur, K. C., & Lamberson, L. R. (1977). *Reliability in engineering design*. New York, NY: Wiley.

Kececioglu, D. B., & Sun, F. (2003). *Environmental stress screening: Its quantification, optimization and management*. Lancaster, PA: DEStech Publications.

Kielpinski, T. J., & Nelson, W. (1975). Optimum censored accelerated life tests for normal and lognormal life distributions. *IEEE Transactions on Reliability*, *24*(5), 310–320.

Klyatis, L. M., & Klyatis, E. L. (2002). *Successful accelerated testing (Part 1): Strategy, step-by-step, vibration and corrosion testing*. New York, NY: Mir Collection.

Krishnamoorthi, K. S. (1992). *Reliability methods for engineers*. Milwaukee, WI: ASQ Quality Press.

Kroo, I. M. (1997). Multidisciplinary optimization applications in preliminary design - Status and directions (AIAA 97-1408). *In Proceedings of 38th AIAA/ASME/ASCE/AHS/ASC "Structures, structural dynamics and materials" conference, and held in Kissimmee, FL, April 7–10, 1997*. Washington, DC: American Institute of Aeronautics and Astronautics.

Kroo, I. M., Altus, S., Braun, R. D., Gage, P., & Sobieski, I. (1994). Multidisciplinary optimization methods for aircraft preliminary design (AIAA 94-4325). *In Proceedings of 5th AIAA/USAF/NASA/ISSMO symposium on "Multidisciplinary analysis and optimization", and held in Panama City Beach, FL, September 7–9, 1994 (Part 1)* (pp. 697–707). Washington, DC: American Institute of Aeronautics and Astronautics.

Lall, P., Pecht, M., & Hakim, E. B. (1997). *Influence of temperature on microelectronics and system reliability*. Boca Raton, FL: CRC Press.

Lewis, E. E. (1996a). *Introduction to reliability engineering* (2nd ed.). New York, NY: Wiley.

Lewis, K. E. (1996b). *An algorithm for integrated subsystem embodiment and system synthesis.* (Dissertation): Georgia Institute of Technology.

Lewis, K., & Mistree, F. (1997). Modeling the interactions in multidisciplinary design: A game-theoretic approach. *AIAA Journal of Aircraft, 35*(8), 1387–1392.

Luce, R. D., & Raiffa, H. (1957). *Games and decisions: Introduction and critical survey.* New York, NY: Wiley.

Meeker, W. Q., & Hahn, G. J. (1985). *How to plan an accelerated life test: Some practical guidelines.* Milwaukee, WI: ASQ Quality Press.

Meeker, W. Q., & Nelson, W. (1975). Optimum accelerated life-tests for the Weibull and extreme value distributions. *IEEE Transactions on Reliability, 24*(5), 321–332.

Menard, S. W. (2010). *Logistic regression: From introductory to advanced concepts and applications.* Thousand Oaks, CA: Sage Publications.

Menq, C. H., Yau, H. T., Lai, G. Y., & Miller, R. A. (1990). *Statistical evaluation of form tolerances using discrete measurement data.* Dallas, TX: ASME (Winter Annual Meeting 1990).

Metropolis, N., & Ulam, S. (1949). The Monte Carlo method. *Journal of the American Statistical Association, 44*(247), 335–341.

MIL-HDBK-189C. (2011). *Handbook: Reliability growth management.* Washington, DC: U.S. Department of Defense.

MIL-HDBK-781A. (1996). *Reliability test methods, plans and environments for engineering, development qualification and production.* Washington, DC: U.S. Department of Defense.

MIL-STD-1635. (1978). *Reliability growth testing.* Washington, DC: U.S. Department of Defense.

MIL-STD-810G. (2008). *Test method standard for environmental engineering considerations and laboratory tests.* Washington, DC: U.S. Department of Defense.

Mistree, F., Marinopoulos, S., Jackson, D. M., & Shupe, J. A. (1988). The design of aircraft using the decision support problem technique. *NASA CR 88-4134,* 1–217.

Mistree, F., Hughes, O. F., & Bras, B. A. (1993). The compromise decision support problem and the adaptive linear programming algorithm. In M. P. Kamat (Ed.), *Structural optimization: Status and promise* (pp. 247–286). Washington, DC: American Institute of Aeronautics and Astronautics.

Nash, J. (1951). Non-cooperative games. *Annals of Mathematics, 54*(2), 286–295.

National Institute of Standards and Technology (2012). *NIST/SEMATECH e-handbook of statistical methods.* Retrieved from http://www.itl.nist.gov/div898/handbook.

Nelson, W. (1990). *Accelerated testing: Statistical models, test plans and data analyses.* New York, NY: Wiley.

O'Connor, P. D., & Kleyner, A. (2012). *Practical reliability engineering* (5th ed.). Chichester, UK: Wiley.

Parkinson, A., Sorenson, C., & Pourhassan, N. (1993). A general approach for robust optimal design. *ASME Journal of Mechanical Design, 115*(1), 74–80.

Phadke, M. S. (1989). *Quality engineering using robust design.* Englewood Cliffs, NJ: Prentice Hall.

Rao, J. R., Badhrinath, K., Pakala, R., & Mistree, F. (1997). A study of optimal design under conflict using models of multi-player games. *Engineering Optimization, 28*(1–2), 63–94.

Reddy, R., Smith, W. F., Mistree, F., Bras, B. A., Chen, W., Malhotra, A., … Lewis, K. E. (1996). *DSIDES user manual*. Atlanta, GA: Georgia Institute of Technology.

Scholz, F. W. (1995a). Tolerance stack analysis methods. *Research and Technology, Boeing Support and Information Services, 30*.

Scholz, F. W. (1995b). *Tolerance stack analysis methods* (ISSTECH-95-030). Seattle, WA: Boeing Information & Support Services.

Shooman, M. L. (1990). *Probabilistic reliability: An engineering approach* (2nd ed.). Malabar, FL: Krieger Publishing.

Sobieszczanski-Sobieski, J. (1989). Optimization by decomposition: A step from hierarchic to non-hierarchic systems. In J. M. Barthelemy (Ed.), *Second NASA/Air Force symposium on "Recent advances in multidisciplinary analysis and optimization"; Proceedings of a symposium cosponsored by NASA Langley Research Center, NASA Lewis Research Center, and the Wright Research Development Center, and held in Hampton, VA, September 28–30, 1988* (pp. 51–78). Washington, DC: NASA.

Society of Automotive Engineers (2000). *Accelerated testing research*. Warrendale, PA: SAE International.

Staudte, R. G., & Sheather, S. J. (1990). *Robust estimation and testing*. New York, NY: Wiley.

Tappeta, R. V., & Renaud, J. E. (1997). Multiobjective collaborative optimization. *Journal of Mechanical Design, 119*(3), 403–411.

Thermotron Industries (Ed.). (1998). *Fundamentals of accelerated stress testing*. Sittingbourne, UK: Thermotron.

Tobias, P. A., & Trindade, D. C. (2012). *Applied reliability* (3rd ed.). Boca Raton, FL: CRC Press.

University of Michigan - Industrial Development Division (1991). *Product development study: A key to world class manufacture of automotive bodies*. Ann Arbor, MI: University of Michigan.

Vincent, T. L. (1983). Game theory as a design tool. *Journal of Mechanisms Transmissions and Automation in Design, 105*(2), 165–170.

Vincent, T. L., & Grantham, W. J. (1981). *Optimality in parametric systems*. New York, NY: Wiley.

Walker, N. E. (1998). *The design analysis handbook: A practical guide to design validation* (2nd ed.). Boston, MA: Newnes.

Young, W. R. (1990). Accelerated temperature pharmaceutical product stability determinations. *Drug Development and Industrial Pharmacy, 16*(3), 551–569.

Yu, J. C., & Ishii, K. (1994). Robust design by matching the design with manufacturing variation patterns (69-2). In B. J. Gilmore (Ed.), *20th ASME Design Automation Conference, Minneapolis, MN, September 1994* (pp. 7–14). New York, NY: ASME publications.

Answers to selected exercises

Chapter 2

2. 25.05

3. Probability $(x_1 < X < x_2) = 0.095355$
   Probability $(X < x_1) = 0.566184$
   Probability $(X < x_2) = 0.661539$

4. 23.42

5. 0.4512

6. $\mu = 42.50$
   $\sigma = 40.660$

7. $\mu = 3.124$
   $\sigma = 1.809$

8. $\beta = 0.571$
   $\theta = 52.527$

9. 49.4

10. $\mu = 3.125$
    $\sigma = 3.407$

11. 75.356

12. 0.2098

13. 315.28

Probabilistic Design for Optimization and Robustness for Engineers, First Edition.
Bryan Dodson, Patrick C. Hammett and René Klerx.
© 2014 John Wiley & Sons, Ltd. Published 2014 by John Wiley & Sons, Ltd.
Companion website: http://www.wiley.com/go/robustness_for_engineers

14. 1.0

15. 0.0001515

Chapter 3

3. 0.598

4. 0.683

5. 4.16

6. 0.33

7. $g(y) = \int \left| \dfrac{z}{y^2} \right| f_V(z) f_R \left(\dfrac{z}{y} \right) dz$

8. $g(y) = \dfrac{1}{y^2 \sigma \sqrt{2\pi}} e^{-\frac{1}{2}\left(\frac{\frac{1}{y}-\mu}{\sigma} \right)^2}$

9. $\mu = 100$
   $\sigma = 4.47$

Chapter 4

1. 2.43

2. 1.31

3. 0.27

4. 26%

5. $\mu = 100$
   $\sigma = 4.5$

6. 1.2%

7.

Chapter 5

1. (a) Nominal life: 22.05, standard deviation: 3.27

   (b) Assuming normal distribution: below = 9.40%, above = 12.88%

2. (a) Lower power than spec: 29.8%

   (b) Theoretical value: 2 400 000
       Actual value: 2 472 000

(c) Log mean parameter = 14.67
Log STD parameter = 0.3096

(d) Below spec: 29.82%
Above spec: 28.87%

3. (a) SD: 1003

(b) SD: 997

(c) Non-symmetrical distributions for input parameters have a small effect on the ability to predict the output standard deviation, usually less than 10% error on the predicted output distribution.

4. (a) 108

(b) 110.4

(c) SD is underestimated by 2.4.

Chapter 6

1. (a) 0.71

(b) 0.67

2. (a) 0.47

(b) The company should improve Y3 since it has the highest weightage for hitting the target and an increase in desirability would be the largest for this requirement. The new product desirability when Y3 hits the target is 0.89.

3. (a) 0.64

(b) **Desirability chart**

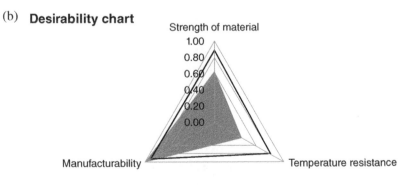

Requirements to be improved
  a   Strength of material
  b   Temp. resistance

4. (a) 0.85

   (b) 34.7, 0.3533

Chapter 7

1. (a) $Y = 60$

   (b) Bias $= 540$; SD $= 12\ 445$

   (c) $Y = 0, Z = 165.48$; SD is 2671

2. $T = 20, A = 5000, C = 1$

3. $A = 551.2, H = 750$

4. (a) $M = 140\ 000$ N, $a = 50$ m/s^2, $A = 926.08$ m

   (b) 1.248

   (c) Lognormal

   (d) 1.15

Chapter 8

1. (a) Approximately 0.91

   (b) E (deflection is the output controlling system C_{pk})

   (c) No, none of the inputs are constrained.

   (d) Reducing the standard deviation of E to 1% of the nominal value of E results in a system C_{pk} of 1.08 before optimization. After optimization, the system C_{pk} is improved to 1.11.

2. Approximately $552.18

3. (a) Approximately 1.57

   (b) f, time $= 5$ is the output controlling system C_{pk}, and f has the largest contribution to the variance at $t = 5$

   (c) Yes, f is constrained at the upper level

   (d) Approximately 1.99

4. Approximately 2350

5. Zero

6. Less than 1

Chapter 9

1. (a) 6.1 ± 2.4

   (b) 6.1 ± 1.2

Chapter 10

1. Factors B and C are significant

2. $C = 17.82, F = 37.78$

3. $C = 23.1, F = 51.6$

4. $B = 48.8, C = 71.9, E = 13.6, J = 10.2$

Chapter 11

1. Answer: 2.53

2. Lubrication is not significant. Set blank holding force < 380

Index

Probabilistic Design for Optimization and Robustness for Engineers, First Edition.
Bryan Dodson, Patrick C. Hammett and René Klerx.
© 2014 John Wiley & Sons, Ltd. Published 2014 by John Wiley & Sons, Ltd.
Companion website: http://www.wiley.com/go/robustness_for_engineers

Printed and bound by CPI Group (UK) Ltd, Croydon, CR0 4YY

17/01/2023

03181289-0003